BTEC First in

Applied Science 2

Rebeka Hasan-Boolaky
Michelle Moran
Series Editor: Nigel Heslop

HODDER
EDUCATION
AN HACHETTE UK COMPANY

Risk Assessment
As a service to our users, a risk assessment for this text has been carried out by CLEAPSS and is available on request to the publishers. However, the publishers accept no legal responsibility on any issue arising from this risk assessment. Whilst every effort has been made to check the instructions of practical work in this book, it is still the duty and legal obligation of schools to carry out their own risk assessments.

Orders: please contact Bookpoint Ltd, 130 Milton Park, Abingdon, Oxon OX14 4SB. Telephone: (44) 01235 827720. Fax: (44) 01235 400454. Lines are open 9.00–17.00, Monday to Saturday, with a 24-hour message answering service. Visit out website at www.hoddereducation.co.uk

First published in 2011 by
Hodder Education
An Hachette UK Company,
338 Euston Road
London NW1 3BH

Impression number 5 4 3 2 1
Year 2015 2014 2013 2012 2011

Cover photo © TEK IMAGE/Science Photo Library

Illustrations by Barking Dog Art

Typeset in Univers 11/13pt by DC Graphic Design Ltd, Swanley Village, Kent

A catalogue record for this title is available from the British Library

ISBN 978 1 444 112092

Contents

Unit 4

Chapter 1
Bonding in substances

Helium party balloons!

Q: Why do we use helium in party balloons when hydrogen is much lighter?

A: It's because hydrogen is flammable and helium is not.

A: Helium is a noble gas and has a full outer shell of electrons, this makes it very unreactive and safe to use.

A: When elements bond together to make compounds they can gain, lose or share electrons in order to have a full outer shell. This makes them more stable.

Figure 1.1

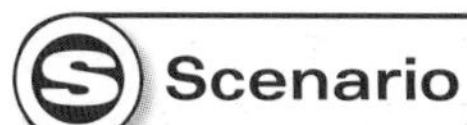

Scenario

You are working for a chemical company that makes different types of chemical substances for different uses. The substances are used in homes, schools, industry and medical care.

Your job is to identify the type of bonding present in each of the substances so that people can decide how the substances can be stored safely.

You need to carry out experiments and physical tests to identify the bonding types and describe the properties of each substance. You may also have to explain why each substance has a different bonding type. You must use the results of your experiments and tests to provide evidence for each type of bonding.

Grading criteria for Bonding in substances

To achieve a pass grade you have to:	To achieve a merit grade you also need to:	To achieve a distinction grade you also need to:
P1 carry out experiments to identify compounds with different bonding types	M1 describe the properties of chemical substances with different types of bonds	D1 explain why chemical substances with different bonds have different properties

Scenario

You are a forensic scientist. You have to identify some unknown substances from a crime scene. You must have a detailed knowledge of how evidence is gathered at the crime scene without causing contamination. To do this you need to know about handling chemicals.

Back in the lab, you will then be testing the samples you have gathered to find their physical and chemical properties. You will make standardised physical and chemical measurements on the materials to obtain a large amount of detailed data to process. The data needs to be gathered with great care as it may be used in legal proceedings.

The first step in your approved procedure is to find out what type of bonding is in the different substances. Using this information you must then decide what type of analysis technique to use. You must know about substances and bonding or you may waste all your evidence samples.

Keeping it local

- Petrochemicals companies have local laboratories that check the purity of the fuels and products they supply.
- Building suppliers need to know the performance of the materials they are using. This requires the materials to be analysed carefully and regularly.
- Makers of medicines need to be very careful about the materials that are in their drugs. The drugs must be tested carefully.

Key words:

boiling point, brittle, conduct, covalent, delocalised, dissolve, ductile, electricity, electron, hard, ion, ionic, lattice, malleable, melting point, metallic, shiny, strong

Definitions activity

Write a scientific definition for each key word. These will be really useful as you work your way through the chapter.

Careers

- Fuel systems technician
- Forensic science officer
- Architectural technician
- Medical technician
- Medicines nurse
- Pharmacy worker
- Building materials technician
- Quality control technician

Background science for the assignments

When atoms have a full outer shell of **electrons** they become very stable and unreactive. Like the noble gases in group 0 of the Periodic Table, they have practically no chemical reactions at all. But most atoms do not normally have a full outer shell. Atoms react in order to achieve a more stable arrangement of electrons. They achieve this by either:

- sharing electrons, which is called **covalent bonding**
- transferring electrons, which is called **ionic bonding**
- having 'loose' electrons moving between many atoms, which is called **metallic bonding** and occurs in metals.

Covalent bonding

When non-metals react together they share electrons to form molecules or giant structures.

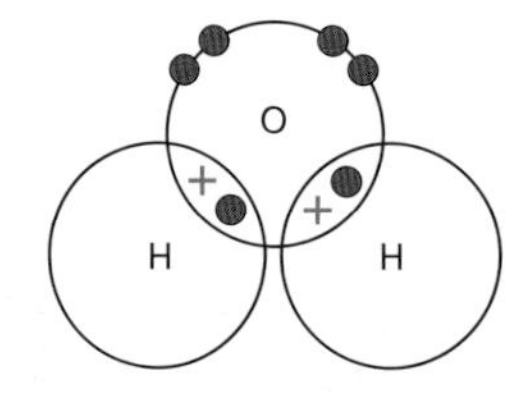

Figure 1.2 An oxygen atom and two hydrogen atoms share their outer electrons to form a covalent bond, which gives the water molecule a stable arrangement.

Ionic bonding

When metals and non-metals react together they usually form substances held together by ionic bonding.

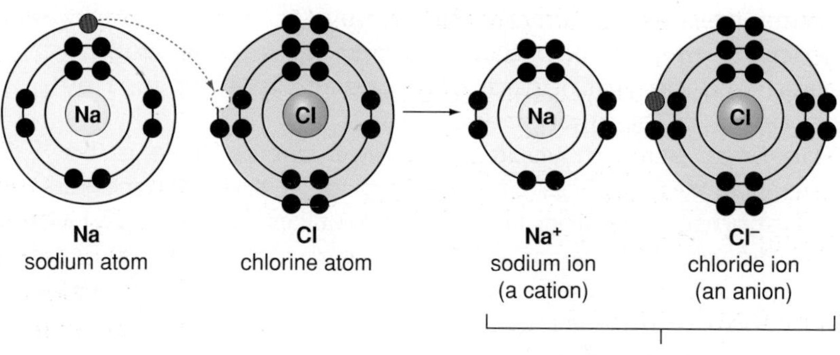

Figure 1.3 When sodium reacts with chlorine it forms an ionic bond. Sodium (metal) loses an electron and chlorine (non-metal) gains an electron.

Metallic bonding

The bonding in metals is a result of positively charged metal atoms arranged in regular layers. The outer electrons in each atom move from one atom to the next causing a sea of 'free' electrons. These electrons are attracted to the positively charged metal ions and bond them together.

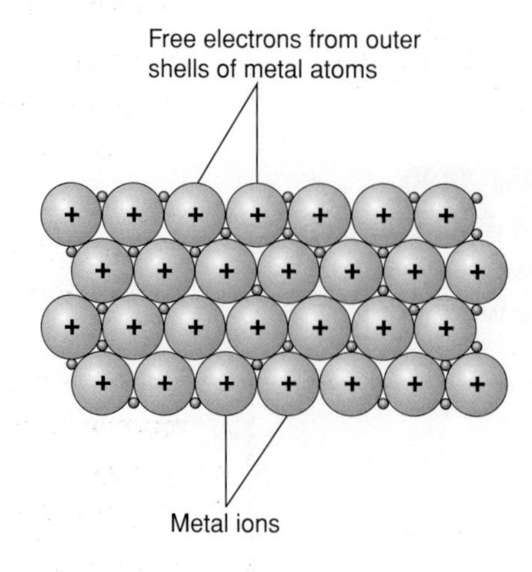

Figure 1.4 Metallic bonding occurs in metals. A sea of 'free' electrons are attracted to the positively charged metal atoms.

Key points to remember about bonding

Covalent substances

- Covalent substances are formed when atoms share electrons.
- The atoms are then held together because they are sharing electrons.
- Covalent substances are formed between non-metal atoms.
- Properties include: low **melting point**, low **boiling point**, do not **conduct electricity**. (There is a giant covalent compound called graphite that does conduct electricity due to its free electrons, which are not involved in the bonding process. However, this is an oddity that does not fit with the pattern.)
- A more normal giant structure is silicon dioxide (sand).

Ionic substances

- Ionic substances are formed when atoms transfer electrons.
- The atom that loses an electron becomes a positive ion and the atom that gains an electron becomes a negative ion.
- The atoms are held together by strong forces between positively and negatively charged ions.
- Ionic compounds are formed between metals and non-metals.
- Properties include: high melting points, high boiling points, conduct electricity when molten or **dissolved** in water.

Metallic substances

- The atoms in metallic substances are arranged in layers.
- The positive metal ions are held together by the sea of 'free' electrons in the outer shell of the metal atoms.
- Properties include: conduct heat and electricity; shiny (have lustre); very strong and **hard**; **malleable** and **ductile** (bend without breaking).

Key Words:

boiling point, bond, conduct, covalent, dissolve, ductile, electricity, electron, hard, ion, ionic, malleable, melting point, metallic

Bonding activities

1 Copy and complete the following sentences by filling in the blanks using the following words:

shared giant covalent molecules

When non-metal atoms react with each other they form __________ substances. The atoms are held together in these substances by __________ electrons. Many covalent substances have particles that are small __________, but a few have __________ structures extending for millions of atoms.

2 Draw diagrams to show the bonding between the following atoms:

a) two hydrogen atoms

b) a hydrogen and a chlorine atom

c) two hydrogen atoms and an oxygen atom

d) a sodium and a chlorine atom.

3 Which of the substances you have drawn in question 2 are covalent and which are ionic?

4 What are the charges on the atoms in metallic bonding?

5 a) Why do elements in group 1 and group 7 bond easily with each other?

b) What type of bonding is this called?

Identifying substances

Essential science for P1

Different substances have different types of bonding. You can identify the type of substance by looking at its physical and chemical properties. Some of its properties can easily be seen, e.g. most metals are shiny and hard, but other properties need to be investigated using **experimental methods**.

Experimental methods to investigate bonding in substances

- Appearance: look at the substances through a hand lens. Can you see any crystal structures? Look for lots of sharp corners and flat edges and sides. Look for similar shapes of these particles.
- Solubility: does a small amount of the substance dissolve when it is shaken up in water?
- Hardness: scratch the substances with a mounted needle or craft-knife blade and view the scratched area using a hand lens. Has the substance scratched easily (soft)? Or is it difficult to scratch (hard)?
- Melting point: put half a spatula of the substance into a boiling tube and hold it using a test tube holder above the blue cone part of a Bunsen burner flame. Does it melt easily?
- Electrical conductivity: put a small amount of the substance into a beaker. Lower carbon graphite electrodes (which are connected to a bulb and power supply) into the beaker (Figure 1.5). Slowly add water from a wash bottle and observe what happens. You may need to swirl the beaker to encourage the substance to dissolve.

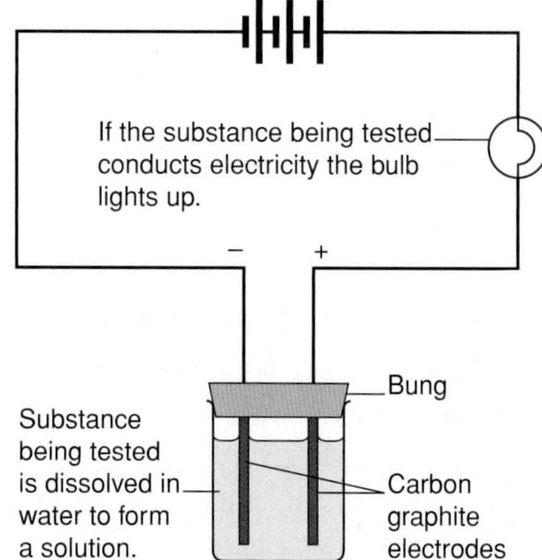

Figure 1.5 Testing a substance for electrical conductivity.

Table 1.1 shows some of the properties that covalent, ionic and metallic substances may have. (The information in this table will be useful for criteria M1 and D1 also.)

Property	Covalent	Giant covalent	Ionic	Metallic
Appearance and state of matter	usually gas and liquid at room temperature	solid	crystal structure; solid at room temperature	shiny appearance; solid at room temperature (exception is mercury which is a liquid)
Solubility	does not dissolve easily	does not dissolve	many dissolve in water	does not dissolve in water
Hardness	not hard (can be brittle)	hard	hard	very hard
Melting point	low – due to weak intermolecular forces	high	high	high
Boiling point	low – due to weak intermolecular forces	high	high	high
Electrical conductivity	cannot conduct as no overall charge	diamond does not conduct; but graphite does due to its layered structure and free electrons	conducts electricity if dissolved in water or if molten because the ions can move freely	conducts electricity due to free electrons which can move through the layers

Table 1.1 A summary of some of the properties of covalent, ionic and metallic substances.

Task

You have been given some different compounds and you need to work out what type of bonding each substance has.

Your task:

You need to carry out experiments to identify some compounds.

The compounds you may investigate include:

- **sodium chloride (common salt)**
- **ethanol**
- **solid wax pieces**
- **iron filings**
- **stearic acid**
- **potassium iodide**
- **granulated zinc pieces**
- **graphite**
- **silicon dioxide (sand).**

Take care when handling these compounds. Ethanol is highly flammable and stearic acid is an irritant. Make sure you wear eye protection.

Step 1: Collect the compounds

Your teacher will provide you with some substances which will be similar to those listed above.

Step 2: Test the compounds

For each compound test for:

- solubility
- hardness
- melting point
- electrical conductivity.

Step 3: Present your results

You may need to draw a table like this to present your results.

Compound	Does it dissolve?	Is it hard?	Does it have a high or low melting point?	Does it conduct electricity? Yes, when a solid = metal Yes, when dissolved = ionic	Type of bonding the compound has

Step 4: Present your findings

For each compound you test say what type of bonding it has.

Make the grade

P1 For P1, you must complete this task:

Carry out experiments to identify compounds with different bonding types.

Properties of compounds with different types of bonding

Essential science for M1

More about covalent bonding

Covalent bonding is about atoms sharing electrons to get a stable outer shell. This results in little groups of atoms (molecules) held together or big networks of atoms (giant structures) sharing electrons with their neighbours. And this results in two families of covalent compounds with very different properties

Molecular substances such as water, hydrogen, oxygen, paraffin wax and carbon dioxide melt and boil easily – although some molecular chemicals such as sugar (sucrose) will decompose when they are heated. All the gases, liquids or soft solids at room temperature are 'molecular'. These molecules can dissolve in liquids like water or other solvents. Oxygen gas and carbon dioxide both dissolve in water. Molecular substances will not conduct electricity either as solids or when dissolved.

Covalent bonds involve two atoms each putting one electron into a shared pair. In some compounds, the sharing is not equal. This results in a dative covalent bond. For example in carbon monoxide the carbon atom shares two electrons with the oxygen atom, but receives a share of **four** electrons from the oxygen. This results in a full outer shell for all the atoms.

More about ionic bonding

Ionic bonding is all about transfer of electrons.

Examples of ionic bonding are: sodium chloride (NaCl), magnesium chloride ($MgCl_2$) and all the salts. The transferred electrons mean the atoms are separate charged ions. The ions are able to conduct electricity when molten or dissolved in water.

Ionic compounds have high boiling and melting points.

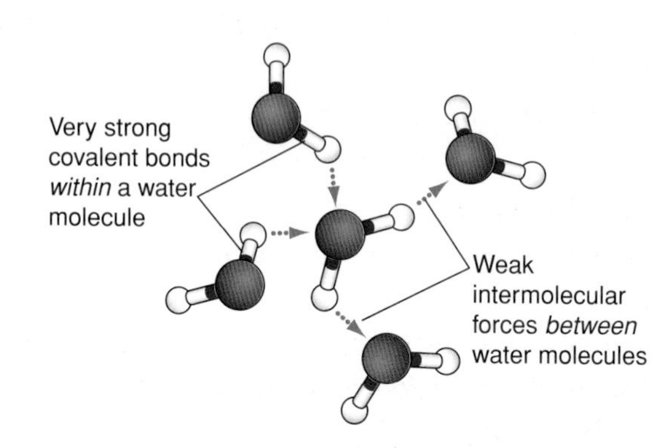

Figure 1.6 Covalent bonding *within* a water molecule is very strong. Intermolecular forces *between* the water molecules are weak.

More about metallic bonding

Metallic substances are usually hard, shiny and can conduct heat and electricity. They can be bent and shaped easily (malleable) and can be pulled into long wires (ductile). The shared free electrons fill the outer energy level and hold the metal structure together. The metals do not dissolve – except in other metals that have free electrons.

Figure 1.7 Copper is ductile – the atoms can slip past each other – so it is used to make flexible electrical wires inside cables.

Key words:

free electrons, intermolecular, ions, molecule

Make the grade

M1 For M1, you must complete this task:

Describe the properties of chemical substances with different types of bonds.

Task

Your task:

You have already identified the type of bonding in some substances for P1.

Now you must produce a presentation to describe the properties that the chemical substances have.

Step 1: Types of bonding

You have identified the type of bonding in some substances by carrying out experiments.

You need to describe the bonding for one of each type of covalent molecule, giant covalent structure, ionic and metallic.

Step 2: Describe the bonding

For each type of bonding, describe the properties of the substances with this type of bonding. You will need to use some of the information that you have gathered from P1. The properties you may wish to describe are:

- **solubility**
- **hardness**
- **melting points**
- **boiling points**
- **electrical conductivity**
- **other properties that you have found.**

Step 3: Present your results

For each type of bonding you need to write about the properties substances have. You will also need to draw a dot and cross diagram to show that you understand the type of bonding it has. The table below may help you to plan your presentation.

Type of bonding	Dot and cross diagram to show bonding (name the atoms involved)	Properties (write some information about the properties the substance has)
covalent molecules, e.g. ethanol		
giant covalent, e.g. silicon dioxide		
ionic, e.g. sodium chloride		
metallic, e.g. zinc		

Step 4: Present your findings

Present your findings in a clear format. You may wish to prepare a PowerPoint presentation or a poster.

Why do different types of bonding have different properties?

Essential science for D1

Covalent molecular substances

Many covalent substances form separate molecules in which the atoms are bound tightly to one another. Unlike in **giant structures** (ionic and covalent) these molecules do not interact with each other much (except through relatively weak forces called intermolecular forces) which makes them very easy to pull apart from each other. Since they are easy to separate, covalent molecular compounds have low melting points and low boiling points.

Giant covalent structures

In some covalent substances the bonds link atoms together in a giant 3-D **lattice**. An example of this is silicon dioxide where the covalent bonds bind each silicon atom to four oxygen atoms and each oxygen atom to two silicon atoms. This links everything together in a strong giant structure. Silicon dioxide is not reactive, it is hard and insoluble and it does not conduct electricity.

Strong covalent bonds between the carbon atoms in each layer

Weaker forces hold the layers together

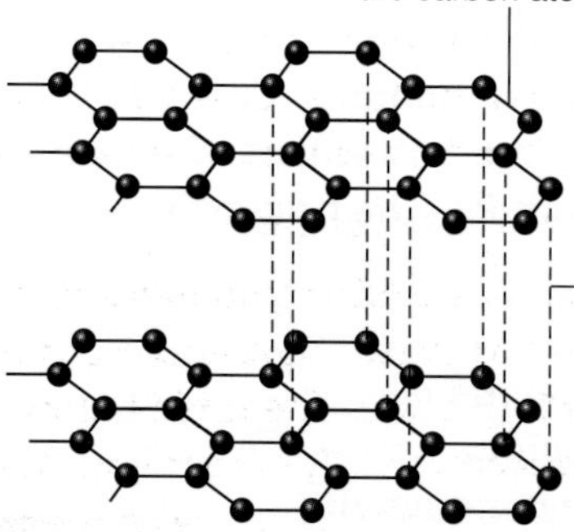

Figure 1.8 Graphite has a giant covalent structure. It is unusual in that it does conduct electricity.

Ionic substances

Ionic substances have high melting points and high boiling points because the electrostatic attraction between positive and negative ions is strong. It takes a lot of energy to overcome this attraction in order to allow the ions to move more freely and form a liquid. Solid ionic compounds do not conduct electricity because the ions are locked into a rigid lattice structure. However, when molten, the ions are free to move out of the lattice structure and carry an electrical charge. Ionic solids are brittle (break easily).

Metallic bonding

The bonding in metals is a result of positively charged metal atoms arranged in regular layers. The outer electrons in each atom can actually move freely from one atom to the next. This is like a sea of 'free' electrons surrounding the rest of the metal atoms (see Figure 1.4 on page 4). Electricity is the movement of electrons, so this is the reason why metals can conduct electricity when attached to an energy source.

Key words:

giant structure, lattice

Make the grade

D1 For D1, you must complete this task:

Explain why chemical substances with different bonds have different properties.

Task

Your task

You have already identified the type of bonding in some compounds and described their properties.

Now you must produce a presentation to explain their properties.

Step 1: Explain the bonding

For each type of bonding, explain why the substances have these properties. You will need to use some of the information that you have gathered from pages 7 and 9. The properties you need to explain are:

- **solubility (Is it soluble? If so, why?)**
- **hardness (Is it hard? If so, why?)**
- **melting point (Why does it have a high or low melting point?)**
- **boiling point (Why does it have a high or low boiling point?)**
- **electrical conductivity (Can it conduct electricity? If so, why?)**
- **any other properties that you have found (brittle, malleable, ductile, shiny etc.).**

Step 2: Present your results

For each type of bonding you need to write about the properties they have. You will also need to draw a dot and cross diagram to show that you understand the type of bonding it has. The table below may help you to plan your presentation (you can get a copy of this from your teacher).

Type of bonding	Properties	Why it has this property
covalent, e.g. ethanol or wax	solubility hardness melting point boiling point electrical conductivity	
giant covalent, e.g. silicon dioxide	solubility hardness melting point boiling point electrical conductivity	
ionic, e.g. sodium chloride	solubility hardness melting point boiling point electrical conductivity	
metallic, e.g. copper or zinc	as well as those above, include also ductile, malleable, shiny, reflective properties	

Step 4: Present your findings

Present your findings in a clear format. You may wish to prepare a PowerPoint presentation or a poster.

Unit 4

Chapter 2
Energy changes

Hot or cold?

Figure 2.1 Self-heating cans are good for making hot drinks on cold days, but self-cooling cans are not available.

All chemical reactions involve energy transfers – that's what makes them happen. Using food for energy in your body and burning fuels in power stations are similar reactions involving oxygen. These chemical systems transfer energy out, causing heat. Other common reactions, like baking bread, take energy in to make the changes happen. In this section we will look at small self-contained packs of chemicals that react to transfer energy.

Scenario

There are a number of self-heating cans available for people to have hot coffee or hot soup. On hot days, it would be nice to be able to have a self-cooling can, but there are none currently available. Your task is to work out how self-heating cans work and then design a self-cooling can which you could pitch to a soft-drinks company. You will need to select the best chemicals to use in the cans and explain how both types work to heat or cool drinks.

Your design needs to include details about how much drink can be heated, the temperature rise of the drink (so it is not too hot causing burns to customers) and the amount of chemicals you need to build into the self-heating can. You should include costings if possible.

You also need to include similar details for your self-cooling can – the risk of scalds is not there, but if your drink freezes ... it is not a drink! You should include costings if you can.

Grading criteria for Hot or cold

To achieve a pass grade you need to:	To achieve a merit grade you also need to:	To achieve a distinction grade you also need to:
 P2 carry out experiments to investigate exothermic and endothermic reactions	M2 explain the temperature changes that occur during exothermic and endothermic reactions	D2 explain the energy changes that take place during exothermic and endothermic reactions

Scenario

Astronauts need to take food on their missions that is easy to prepare inside the spacecraft. You have been asked to design the packaging for a range of ready-made meals which can be 'self-heated' or 'self-cooled'. You need to explain to the astronauts how the reactions in the packaging work.

Figure 2.2 Astronauts need to be able to prepare 'self-heating' or 'self-cooling' meals easily when inside their spacecraft. Athletes use 'self-cooling' ice packs when they get injuries.

Keeping it local

- Do you play any sports? You could talk to a coach at your local sports centre about the use of cooling packs and heat packs to treat injuries.
- Do you live near an army centre? You could arrange to visit the centre and ask if they use self-heating food packs when they are on long exercises.

Careers

- Sports therapist
- Paramedic
- Mountain rescue officer
- Fast-food retailer
- Soft-drinks company designer
- Space science company designer
- Outdoor survival expert
- Tour organiser to far-away places

Background science for the assignments

Chemical reactions and energy

It is not always obvious that a chemical reaction is happening. However, a chemical reaction always involves an energy transfer. So, a good way to see if a reaction is happening is to see if there is a temperature change. Temperature is a measure of how much energy transfer has happened.

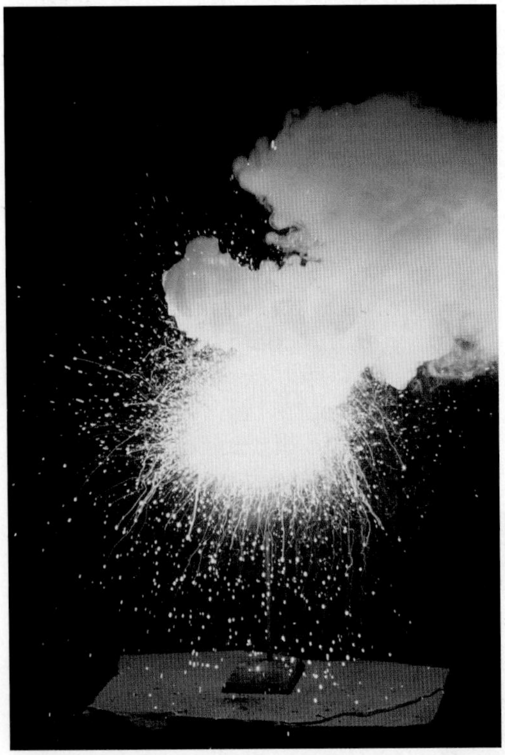

Figure 2.3 It can be very obvious that a chemical reaction is happening. This picture is of the thermite reaction between aluminium and iron oxide. As well as producing iron, it releases a lot of energy as heat and light.

A thermometer is used to measure temperature changes. There are different types of thermometer; alcohol thermometers are not precise and it is easy to read them incorrectly. Digital thermometers are more precise.

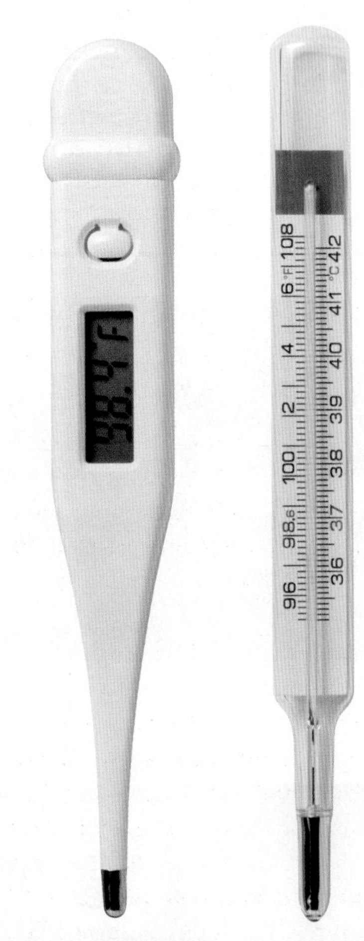

Figure 2.4 Thermometers such as the mercury thermometer (right) and the digital thermometer (left) are used to measure changes in temperature.

When science **investigations** are carried out care must be taken to collect valid and **reproducible** evidence. This can then be used to make a decision or a hypothesis. Using the best equipment to collect results helps to give us good evidence.

Collecting good evidence requires knowing about input variables, output variables and control variables.

Key words:

control variable, hypothesis, input variable, investigation, output variable, reproducible

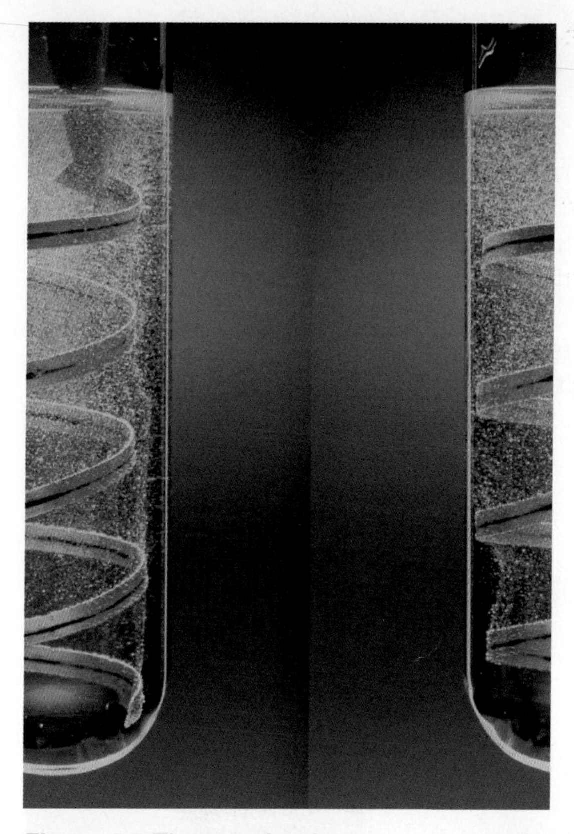

Figure 2.5 The reaction between magnesium ribbon and hydrochloric acid releases heat. The test tube can get quite hot!

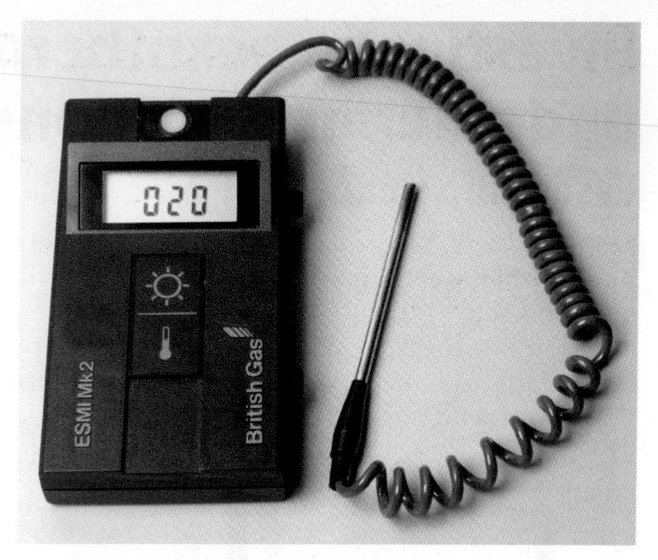

Figure 2.6 A temperature probe can be used to collect information very precisely. The data is logged and can be seen on a computer as a table or graph. As with all equipment, there can be human error so the probes must be used correctly to collect valid results.

Hot or cold activities

1 Look at the key words on page 14. Write a definition for each term.

2 a) Ask your teacher for a few pieces of magnesium ribbon and some hydrochloric acid. Plan a simple investigation to find out what happens to the temperature of the acid when you put different amounts of magnesium ribbon into it. Include all the step-by-step details and any safety precautions you will take. Use a thermometer to measure the temperature changes in the hydrochloric acid.

b) Now use a data-logger in place of the thermometer. Set up the experiment in exactly the same way. Do you get the same result? Which is easier to use? Which is better to use?

3 Plan and carry out an investigation to find out what happens when potassium nitrate is dissolved in water. Include all the step-by-step details of your investigation and any safety precautions you will take. If you are working in a classroom some students should carry out the investigation using a thermometer and some using a data-logger. Compare your results with the rest of the class.

4 Plan and carry out an investigation into which solid fuel will give the most energy. Use a fixed amount of water and burn a known quantity of fuel each time.

Investigating exothermic and endothermic reactions

Essential science for P2

Many chemical reactions give out heat and feel warm. These are called **exothermic reactions**. Heat energy is transferred *from* the system to the surroundings.

In some other types of chemical reactions, the opposite happens and the reaction gets colder. These are called **endothermic reactions**. Heat energy is transferred *into* the system from the surroundings.

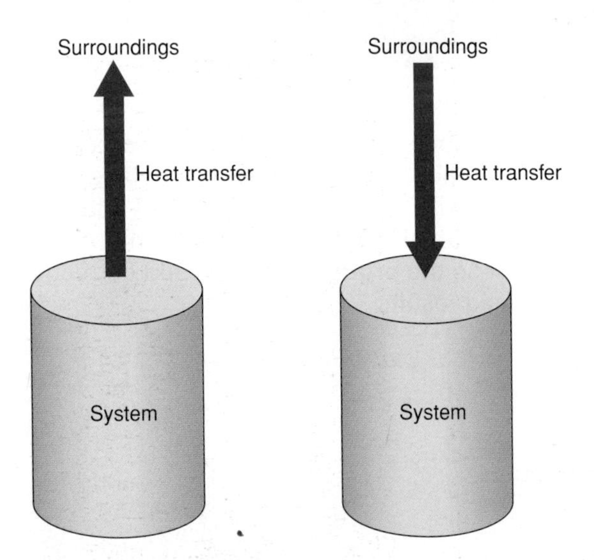

Figure 2.7 In an exothermic reaction (left) heat energy is transferred from the system into the surroundings. In an endothermic reaction (right) heat energy is transferred into the system from the surroundings.

Figure 2.8 Using data-logging equipment to measure the temperature change during a reaction gives more accurate measurements.

When the temperature change in a chemical reaction is investigated it is important to make sure that the results are reliable and accurate.

Things to remember when planning an investigation of this type:

- Take the temperature before and after the reaction to find the overall change in temperature.
- Repeat the test three times to make sure that the results are reliable.
- Keep the control variables the same.
- Insulate the apparatus to prevent heating and cooling from the environment.
- If possible, carry out the reactions in a plastic or polystrene cup supported in a glass beaker to help accuracy.
- Make sure accurate measurements are taken.

Key words:

endothermic reaction, exothermic reaction

Task

Carry out chemical reactions and decide whether they are exothermic or endothermic. You can then decide which reactions would be suitable for a self-heating can and which would be suitable for a self-cooling can. Use the worksheet provided.

Your task:

Step 1: Practical

Follow the practical guidelines on the worksheet to investigate the temperature changes that occur for each chemical reaction. Record your results on the worksheet. State whether each reaction is exothermic or endothermic.

Here are some reactions that your teacher may give you to investigate:

- **mixing fresh calcium oxide and water**
- **dissolving sodium thiosulfate**
- **mixing powdered cement mixture with water**
- **dissolving ammonium nitrate.**

Step 2: Calculations

Calculate the amount of energy transferred in each of your heating reactions. Use this formula:

$$\text{Heating energy (Joules)} = \text{Mass of water heated (Kg)} \times 4.2 \times \text{Change in temperature (°C)}$$

Step 3: Produce a poster of your 'self-cooling' can design

- **Produce a poster to present to a drinks company. The poster needs to include detailed design plans. Labelled drawings and diagrams would be helpful to illustrate your poster.**
- **A cross-section diagram of your can would be helpful (see Figure 2.9). Label each part carefully.**
- **Include on your poster information about: which chemical(s) should be used and why; what safety precautions need to be considered; instructions as to how the can would work.**

Make the grade

P2 For P2, you must complete this task:

Carry out experiments to investigate exothermic and endothermic reactions.

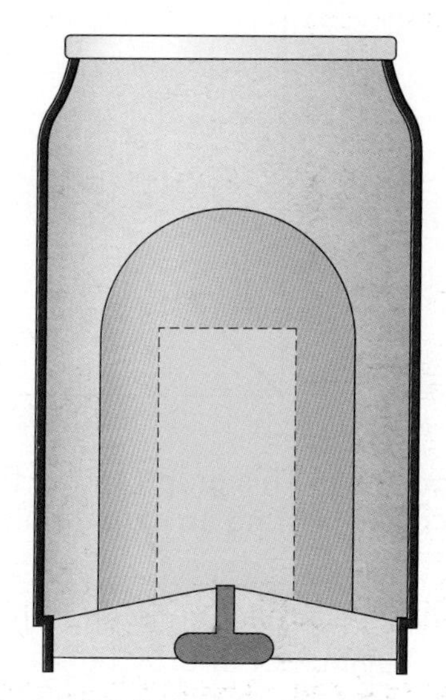

Figure 2.9 Cross section of a self-cooling can

Explaining energy transfers

Essential science for M2

Exothermic reactions

All chemical reactions involve energy transfers. Without an energy transfer nothing would happen.

Chemicals have **potential energy** stored inside them. When this energy is released to the **surroundings** an increase in temperature is detected.

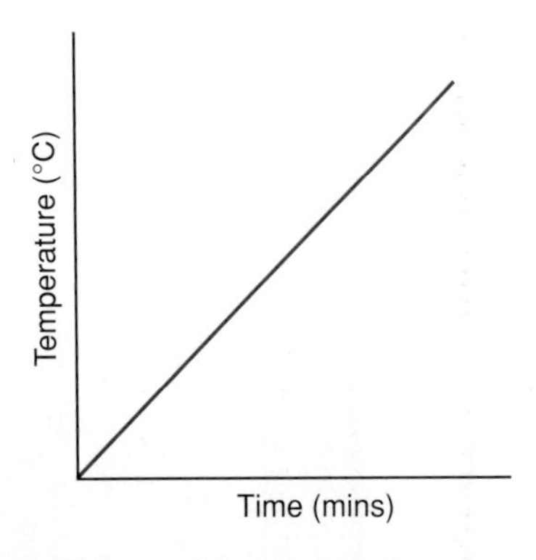

Figure 2.10 This graph shows that temperature increases during an exothermic reaction.

Endothermic reactions

If chemicals do not have enough energy to make a chemical reaction happen they need to take in energy from the surroundings. If energy is taken in and used during a reaction, a decrease in temperature is detected.

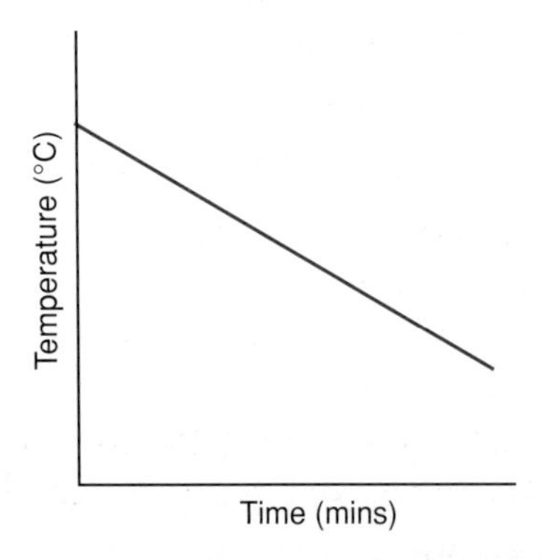

Figure 2.11 This graph shows that temperature decreases during an endothermic reaction.

Whether a reaction is endothermic or exothermic depends on the balance of energy needed to break the chemical bonds in the substances that are reacting (called the **reactants**) and the energy needed to make new chemical bonds in the substances that are made (called the **products**) in the reaction.

Data for chemical reactions often includes information about energy transfer as well as the familiar reaction equation.

For example:

$C_2H_5OH(l) + 3O_2(g) \rightarrow 2CO_2(g) + 3H_2O(l) : \Delta H = -1367$ kJ/mol

This means that when 1 mole (formula mass in grams) of ethanol is burned, 1367 kJ of energy is transferred to the surroundings. A negative value for ΔH, shows an exothermic reaction.

Energy changes for reactions are given as kilojoules per mole (kJ/mol) of reactant.

Key words:

potential energy, product, reactant, surroundings

Task

Develop your self-cooling can proposal to explain what temperature changes happen during the chemical reactions.

You could add the explanations to your design poster or write a speech to 'pitch' your can to a soft-drinks company. Remember to include explanations of the science of how your can works.

Step 1: Cooling reaction

Include the answers to the following questions.

- **Which reaction would you choose to go in a self-cooling can?**
- **Is the reaction exothermic or endothermic?**
- **What happens to the temperature during the reaction?**
- **Why does the temperature change during the reaction?**

Step 2: Heating reaction

Include the answers to the following questions.

- **Which reaction would you definitely *not* choose to go in a self-heating can?**
- **Is the reaction you would choose for your can exothermic or endothermic?**
- **What happens to the temperature during the reaction?**
- **Why does the temperature change during the reaction?**

Step 3: Calculate

Calculate how much energy would have to be transferred to make a can of drink hot (60°C) or cold (4°C). Use the formula:

$$\text{Heating energy (Joules)} = \text{Mass of water (Kg)} \times 4.2 \times \text{Temperature change (°C)}$$

Step 4: Safety instructions

Write a series of instructions for using your self-cooling can. Include information about the hazards associated with the chemicals you are using. Also include information about how cold you think the can will become and how quickly you think this will happen.

When you have finished, show your teacher.

Make the grade

M2 For M2, you must complete this task:

Explain the temperature changes that occur during exothermic and endothermic reactions.

Changing energy levels

Essential science for D2

Did you know?

The Greek letter '**delta**' (shown by a Δ) is used in science to represent change.

In chemical reactions ΔH means 'energy change'.

- A negative energy transfer, $-\Delta H$ means energy stored in the products decreases. This is an exothermic reaction.
- A positive energy transfer, $+\Delta H$ means energy stored in the products increases. This is an endothermic reaction.

You have seen that during a chemical reaction bonds are broken and then new bonds are made. Energy is taken in to start a reaction to break bonds and energy is released when new bonds are made. The overall **change** in temperature for a reaction depends on the amount of bond breaking and bond making that happens.

- If more energy is released due to bond making the temperature increases, energy is given out and the reaction is exothermic.
- If more energy is absorbed to break bonds the temperature decreases, energy is taken in and the reaction is endothermic.

Energy level diagrams are used to show the overall energy change in a chemical reaction (see Figures 2.12 and 2.13). From these diagrams the reaction can be classed as exothermic or endothermic.

Key words:

change, delta, energy level diagram

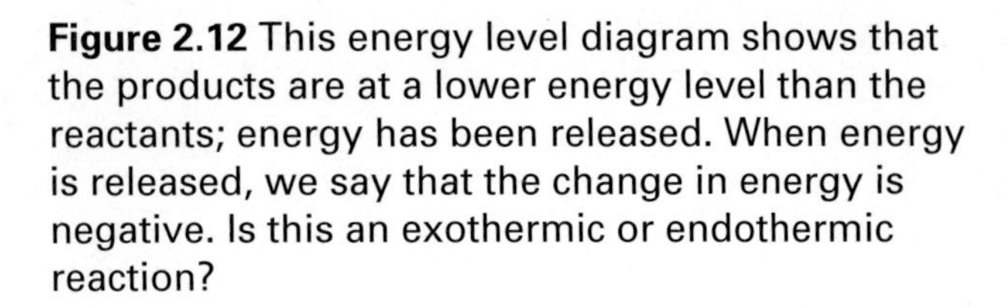

Figure 2.12 This energy level diagram shows that the products are at a lower energy level than the reactants; energy has been released. When energy is released, we say that the change in energy is negative. Is this an exothermic or endothermic reaction?

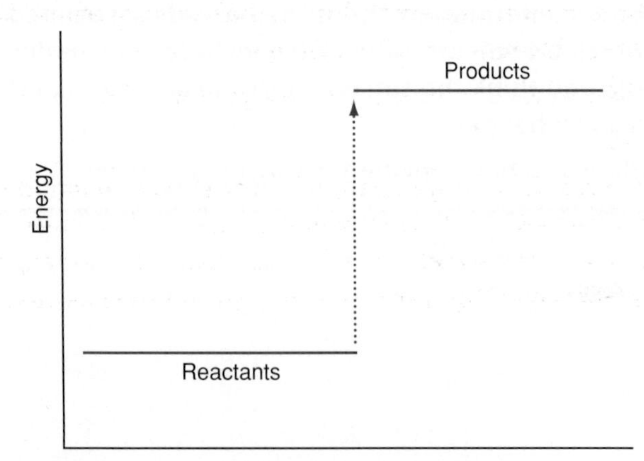

Figure 2.13 This energy level diagram shows that the products are at a higher energy level than the reactants; energy has been absorbed. When energy is absorbed, we say that the change in energy is positive. Is this an exothermic or an endothermic reaction?

Task

Develop your self-cooling can proposal to explain what energy changes happen during the chemical reactions.

You could include the explanations on your design poster or in your pitching speech.

Your task:

Step 1: Cooling reaction

Include the answers to the following questions.

- **Which bonds need to be broken during the reaction?**
- **Which bonds need to be made during the reaction?**
- **Is more energy needed to break the bonds or is more energy released when the bonds are made?**
- **What would the energy level graph look like?**
- **Is the energy change positive or negative?**

Step 2: Heating reaction
Include the answers to the following questions.

- **Which bonds need to be broken in the reaction?**
- **Which bonds need to be made in the reaction?**
- **Is more energy needed to break the bonds or is more energy released when the bonds are made?**
- **What would the energy level graph look like?**
- **Is the energy change positive or negative?**

Step 3: Modelling energy changes

Draw a series of diagrams to illustrate how a molecule changes in a simple reaction.

Show the bonds in the reactant being broken, and show that energy is being taken in at this stage.

Show the atoms reforming to make the product, and show that energy is given out at this stage

When you have finished, show your teacher.

Step 4:

Data for the energy change in a chemical reaction is usually given as 'Kilojoules per mole'. Explain what this is and explain how you would measure and calculate it in an experiment.

Make the grade

D2 For D2, you must complete this task:

Explain the energy changes that take place during exothermic and endothermic reactions.

Unit 4

Chapter 3
Organic compounds

Complex carbon compounds

There are hundreds of everyday products that are produced from organic compounds. Figure 1 shows just a few of them. Organic compounds are based on chains of carbon atoms. They are not necessarily made from living things.

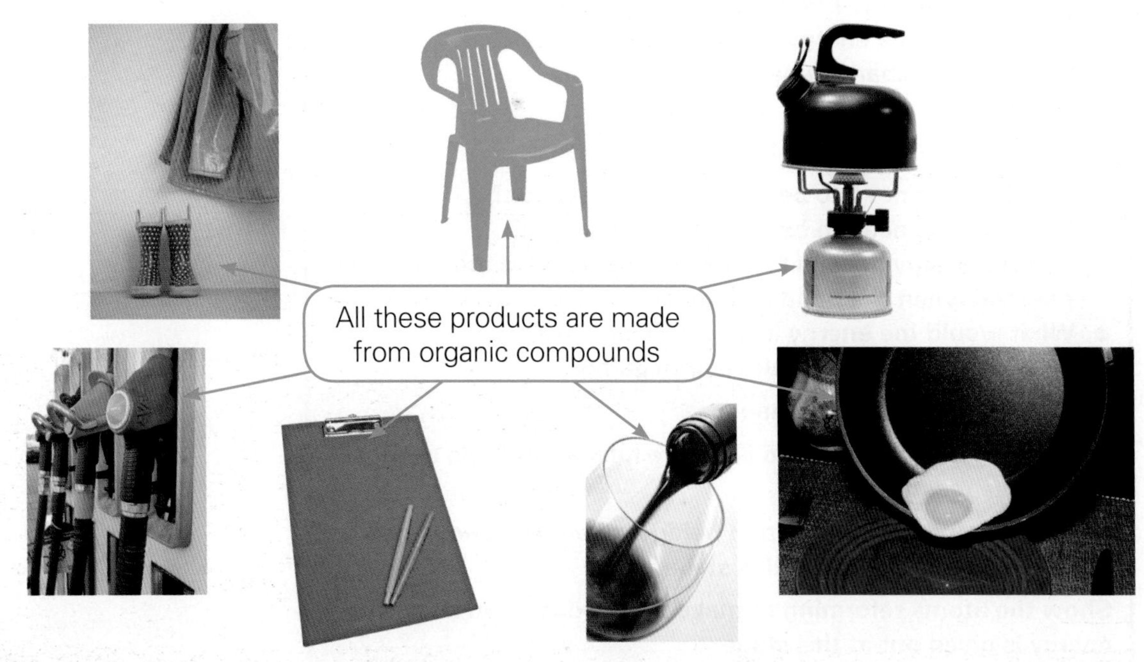

Figure 3.1

Scenario

You are a journalist working for a local television station. You have been asked to produce a series of television programmes to raise awareness about the ways in which organic chemicals are used in everyday life.

You have to explain the nature of these carbon compounds, why they are called 'organic', why there is a whole series of similar compounds and what some of the reactions and uses of these compounds are.

You may need to produce up to three separate TV programmes:

- programme 1: identifying organic compounds
- programme 2: the uses of organic compounds
- programme 3: the benefits and disadvantages of organic compounds.

Grading criteria for Organic compounds

To achieve a pass grade you have to:	To achieve a merit grade you also need to:	To achieve a distinction grade you also need to:
P3 carry out experiments to identify organic compounds	**M3** describe the uses of organic compounds in our society	**D3** explain the benefits and disadvantages of using organic compounds in our society

Scenario

You are a petrochemical engineer who works for a large oil company. Petrochemicals are the main source of organic compounds that we have. They are useful but also often get a bad press and are blamed for pollution.
A local school has asked you to come in and talk to its Year 10 pupils about crude oil. In your presentation you will have to describe some of the useful products that come from crude oil. You should also explain what fractional distillation is.

(a) Millions of years ago, tiny sea plants and animals died and were buried on the sea floor. Over time they were covered by layers of sand and silt.

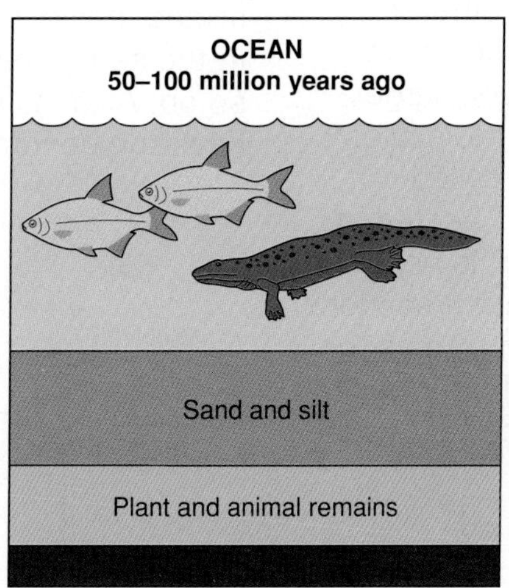

(b) Over millions of years, the remains were buried deeper and deeper. The heat and pressure turned them into crude oil and natural gas.

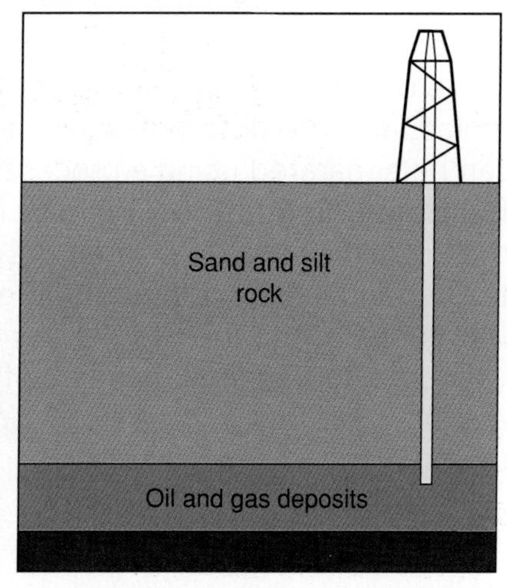

(c) Today we can drill down through the layers of sand, silt and rock to reach the oil and gas deposits.

Figure 3.2 Crude oil formation.

Keeping it local

- Are there any factories near you that manufacture products from organic compounds?
- Where does the petrol at your local petrol station come from?
- Where does the crude oil get refined?

Careers

- Pharmacist
- Paint technician
- Oil refinery technician
- Fabric dye technician
- Organic farmer

Background science for the assignments

In science 'organic' means chemicals that have carbon–carbon bonds often repeated over and over to form a long chain. These chemicals can come from man-made sources or from living things, for example crude oil.

What are organic compounds?

Organic chemistry is the study of the structure, properties, reactions and preparation of carbon-based compounds, which are called **hydrocarbons**. Hydrocarbons are compounds that contain only **carbon** and **hydrogen**. Sometimes these compounds may contain other elements, for example nitrogen, oxygen or even one of the halogens, such as chlorine.

The hydrocarbons we use come from crude oil. Crude oil is a mixture of many different hydrocarbons. Some of these molecules in crude oil are small and some are very large molecules. The different types of molecules can be separated using a process called **fractional distillation**. Crude oil is heated at the bottom of a large column. As it is heated the vapours pass up through different sections. It is cooler at the top of the column, so as the vapours rise they condense at different heights and the hydrocarbons can be tapped off and used.

Key words:

alkane, carbon, ethane, fractional distillation, hydrocarbon, hydrogen, methane, organic chemistry, propane

Did you know?

You will be aware of 'organic' food becoming very popular. The word 'organic' relates to plants and living things that are produced without the use of artificial fertilisers and pesticides. 'Organic' can also be used to indicate living organisms.

Hydrocarbons

In Chapter 1 you learnt that carbon atoms form strong covalent bonds with each other. This allows carbon atoms to form long chains. Hydrogen atoms are attached to the carbon atoms in the chains. The simplest hydrocarbons are called **alkanes**. **Methane**, the simplest alkane, has one carbon atom. **Ethane**, the next in the series, has two, **propane** has three, and so on. Alkanes are a group of hydrocarbons that have the general formula C_nH_{2n+2}.

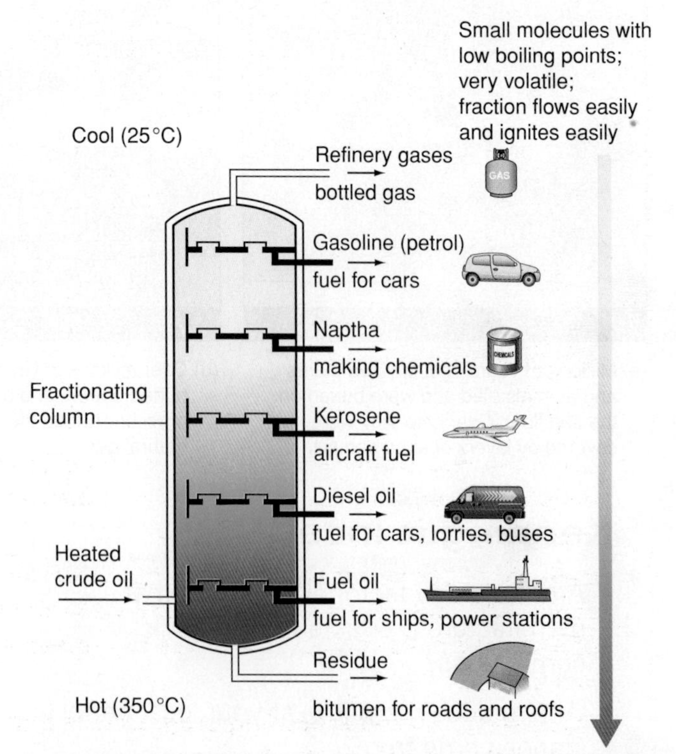

Figure 3.3 The fractional distillation of crude oil relies on different substances in crude oil having different boiling points.

Organic compounds activities

1 Copy and complete the table below or use the alkanes activity sheet.

Name of alkane	Number of carbon atoms	Number of hydrogen atoms	Formula	Structure
methane	1	4	CH_4	H H—C—H H
ethane	2	6	C_2H_6	
propane	3	8		
butane	4			
pentane	5			
hexane		14		
octane	8			
decane		22		

2 Use molymod atomic models, SEP chemical jigsaw or other molecular-model kits to make models of methane, ethane, propane and butane.

If you do not have access to molecular-model kits you could use different coloured Plasticine balls and straws, or polystyrene balls and cocktail sticks to represent the atoms and bonds.

There are ICT-based modelling programmes that your teacher may ask you to use to draw images.

You could make a class display of all your models in your classroom.

Questions

1. What is fractional distillation?
2. Name five products from the fractional distillation of crude oil.
3. Where is the hottest part of a fractional distillation column?
4. What is a hydrocarbon?
5. Where do the smallest hydrocarbon molecules get tapped off in the fractional distillation of crude oil – at the top or at the bottom?
6. Name the first eight alkanes.
7. Why are alkanes useful as fuels?

Identifying organic compounds

Essential science for P3

You need to be able to identify the following five types of organic compounds from their chemical formulae.

1 Alkanes

These have the basic formula C_nH_{2n+2}. Figure 3.4 shows the structure of **butane**, C_4H_{10}.

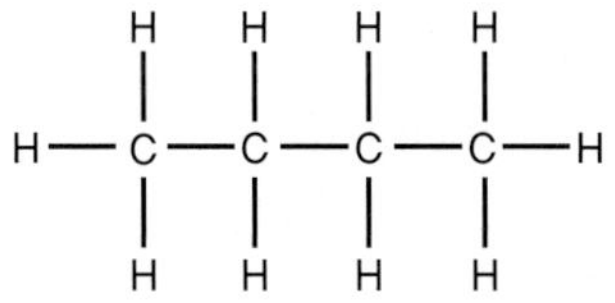

Figure 3.4

2 Alkenes

These have the basic formula C_nH_{2n}. **Alkenes** contain a double C=C bond. Figure 3.5 shows the structure of propene, C_3H_6.

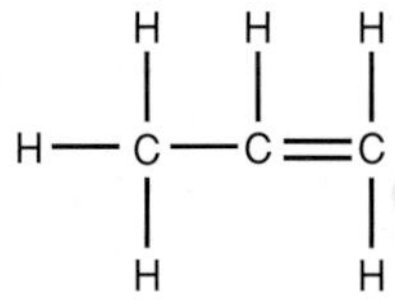

Figure 3.5

3 Alcohols

These have the basic formula $C_nH_{2n+1}OH$. **Alcohols** contain a hydroxy group, –OH. Figure 3.6 shows the structure of ethanol, C_2H_5OH.

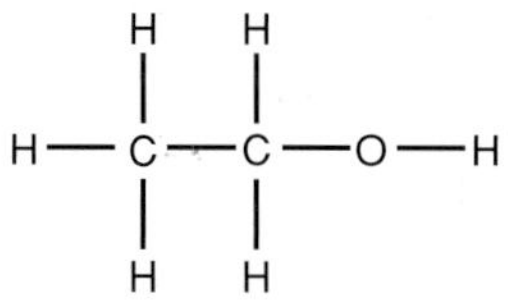

Figure 3.6

4 Carboxylic acids

These contain a carboxyl group, C(=O)OH. Figure 3.7 shows the structures of some **carboxylic acids**. Can you find the carboxyl group in each molecule? Carboxylic acids are weakly acidic. Salicylic acid is related to aspirin and anthranilic acid is a corrosion inhibitor.

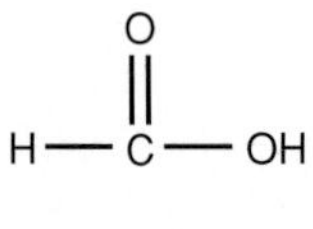

a) Formic acid

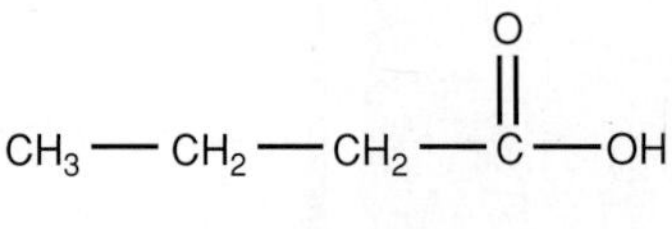

d) Butyric acid

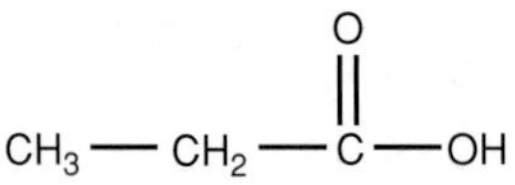

b) Propionic acid

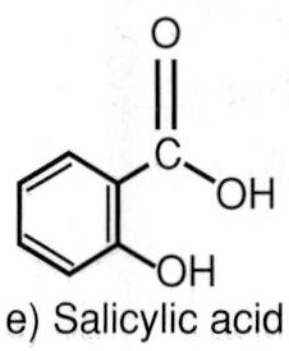

e) Salicylic acid

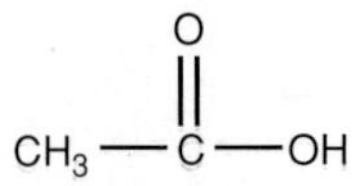

c) Acetic acid
(also known as ethanoic acid)

f) Anthranilic acid

Figure 3.7

5 Contain a halogen atom

Some organic molecules contain a halogen atom, for example bromine (Br) or chlorine (Cl). Figure 3.8 shows the basic structure of chloroethene (vinyl chloride) which can be polymerised to form polychloroethene or PVC (polyvinylchloride).

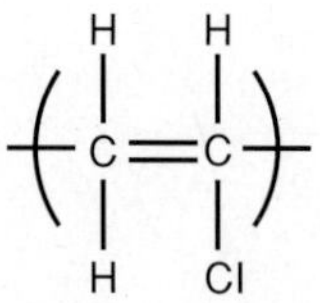

Figure 3.8

Key words:

alcohol, alkanes, alkenes, butane, carboxylic acid, ethane, methane, octane, propane

Task

Your first TV programme is titled 'Identifying organic compounds'.

In this programme you will show the viewers what the five main types of organic compound are, by carrying out experiments that allow you to identify them correctly.

Your task:

Step 1: Collect the organic compounds

Your teacher will provide you with some organic compounds: A, B, C, D and E.

Step 2: Identify the organic compounds

You will carry out practical tests on the organic compounds.

Step 3: Present your results

You may need to draw a table like this to record your results.

Test	Unknown organic compound				
	A	B	C	D	E
Does it react with bromine water?					
Is it a good fuel?					
Is it a good solvent?					
What type of smell does it have?					
What type of flame is produced?					
Does it mix with water?					
Is it tough and durable?					

Step 4: Conclusion

Clearly state which type of organic compound substances A, B, C, D and E are.

Step 5: Present

You may wish to video record a presentation where you identify the compounds, or you could present your work as a report.

Make the grade

P3 For P3, you must complete this task:

Carry out experiments to identify organic compounds.

Uses of organic compounds

Essential science for M3

Organic compounds are used in thousands of everyday products such as furniture, clothing, chemicals, medicines, foods, sports equipment, components for cars, household appliances, computers ... the list is endless.

Crude oil is our main source of fuel and organic chemicals. Crude oil is obtained from places such as the North Sea, the Gulf of Mexico and the Middle East.

One of the main uses of organic compounds is in making **plastics**. The scientific name for plastics is **polymers**. Plastics are made using the **polymerisation** process.

Key words:

crude oil, ethene, plastics, polymerisation, polymers, polythene, thermoplastics, thermosetting polymers

Polymers: Addition polymerisation

Alkenes can be obtained from crude oil. They are very important in the manufacture of plastics and other polymers. Alkenes contain a double bond between two of their carbon atoms (C=C).

Ethene monomers (single molecules) are polymerised under special conditions. These include a high temperature, a high pressure and the presence of a catalyst. The monomers become joined together to form a long polymer ('poly' means many).

In this reaction, many single ethene molecules are polymerised to form poly-ethene (commonly known as **polythene**). Up to 1500 ethene molecules are needed to form one poly-ethene chain.

Other polymers

Other common polymers include polystyrene, PVC and polypropylene. Polystyrene is made from many styrene molecules that have been polymerised. It is a strong, rigid plastic that can also be made into a foam and used as packaging. PVC (polyvinylchloride) is made from vinyl chloride molecules that have been polymerised. Polypropylene (polypropene) is made up of propene molecules.

Some polymer plastics like polyethene soften when they are heated. These are called **thermoplastics**. Others set rigid after they are made, and these are called **thermosetting polymers** or resins. Plastic saucepan handles get hot so they are made from thermosetting polymers rather than thermoplastics.

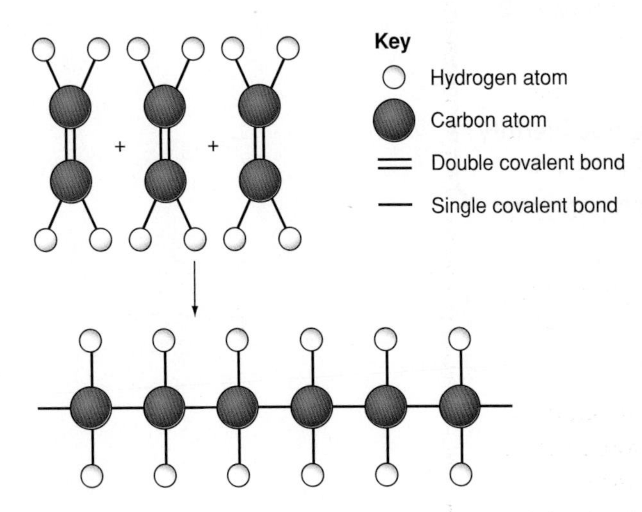

Figure 3.9 Addition of ethene molecules to form polythene.

Figure 3.10 Plastic objects – cheap to make and easy to clean.

Make the grade

M3 For M3, you must complete this task:

Describe the uses of organic compounds in our society.

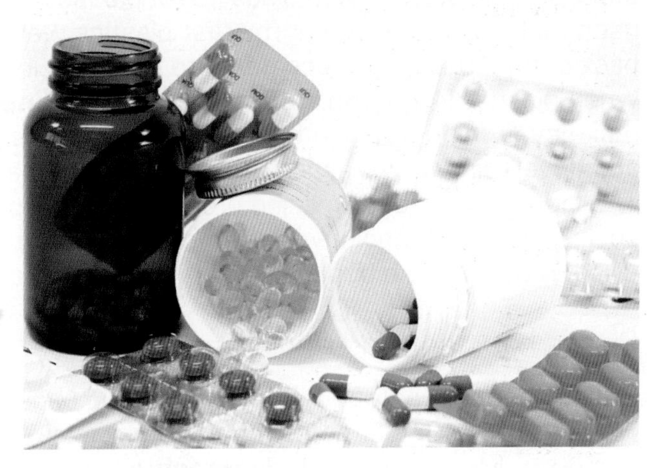

Figure 3.11 Some uses of organic compounds. Beauty products often contain ethanol and esters made from alcohols and carboxylic acids ...

... and medicines contain chemicals made from organic compounds.

Task

Your second TV programme is titled 'The uses of organic compounds'.

In this programme you need to describe some of the uses of different types of organic compound in our society.

Your task:

Step 1: Choose

From your research and prior learning, choose ten different organic compounds.

Step 2: Describe

For each of your ten organic compounds, describe the uses it has in our society today.

Think about the everyday uses of each and how it has changed the way in which we live.

Step 3: Present

You may wish to video record a presentation where you can describe the uses of each of the compounds you have chosen. Or you could present your ideas as a series of cartoons or a comic strip that could be used as part of your TV programme.

When you have finished, show your teacher.

Organic compounds: pros and cons

Essential science for D3

Plastics

Plastics are made from organic compounds (e.g. polythene from ethene). They are useful in the manufacture of millions of different products. In fact, since 1976, plastic has been the most used material in the world! The first plastics were introduced back in the 1920s, but became more common in the 1950s when plastics like **nylon**, **Teflon** and **polystyrene** were used to make everything from clothing to furniture. Since then, making and developing new plastics has grown into a major industry. The industry provides amazing products that help us to live in a society where technology is advancing all the time.

However, the use of plastics has its disadvantages too.

Figure 3.12 Jellyfish or plastic bag?

Plastic bags and the environment

British shoppers get though 8 billion plastic bags a year (equivalent to 133 per person) but the humble plastic bag comes at a price. Most are non-**biodegradable** and can take up to 1000 years to decompose, meaning they end up in the most unusual places and cause all sorts of problems.

In India, cows are eating plastic bags as they forage for food on the street. They then end up choking or starving to death. The same thing happens to turtles, which can mistake plastic bags for jellyfish.

In 2002, Bangladesh put a ban on all polythene bags after they were found to have been the main culprit during the 1988 and 1998 floods that submerged two-thirds of the country. The problem was that plastic bags were choking up the drainage system.

Many other countries have enforced a 'tax' on plastic carrier bags, and this has led to a 97% reduction in their distribution. This is good news for the environment. Do you think the UK should have a tax on plastic bags?

Figure 3.13 Turtles can mistake plastic bags for jellyfish.

The cost of burning fuels

All fuels from crude oil, such as petrol, diesel, kerosene and paraffin, are very high in energy. When they are burnt, a reaction called combustion takes place. Combustion reactions produce carbon dioxide and water. The carbon dioxide produced from combustion gets released into the atmosphere, and this has led to a 'greenhouse effect' and **global warming**.

Key words:

biodegradable, global warming, nylon, plastics, polystyrene, Teflon

Task

Your third TV programme is titled 'The benefits and disadvantages of organic compounds'.

In this programme you need to clearly explain the benefits that organic compounds have in our everyday lives and also the disadvantages of using some of them.

Your task:

Step 1: Benefits

Choose two organic compounds that you have been learning about. Explain the benefits that they have in everyday life. You need to think about the positive effects they have on things like:

- **technology**
- **transport**
- **medical advances and medicines**
- **packaging**
- **chemicals**
- **sport**
- **any other ideas.**

Step 2: Disadvantages

Choose two organic compounds that you have been learning about. Explain the negative effects that they can have on things like:

The environment

- **pollution**
- **global warming**
- **flooding**
- **oil/chemical spills**

Wildlife

- **animals**
- **plants and crops.**

Step 3: Present

You may wish to video record your presentation to use in your TV programme. Or you could present your ideas as a series of cartoons or as comic strip that could be used in the programme.

When you have finished, show your teacher.

Make the grade

D3 For D3, you must complete this task:

Explain the benefits and disadvantages of using organic compounds in our society.

Unit 4

Chapter 4
Special materials

Sporty materials

Figure 4.1

Scenario

Advances in science and technology mean that new materials are being developed all the time that have amazing, unique properties. Many areas of life have benefited from these new materials, including fashion design, medical science, building and technology. The world of sport has benefited too. The new materials have led to the production of sophisticated equipment and clothing that improve performance and have even helped sportsmen and women recover from injury. Again, you are a journalist working for your local television station. Your next television programme is about the different types of new materials used in sport and how the performances of sportspeople can be affected by these new materials. Your task is to investigate the different types of materials and explain to viewers how performance is affected by them.

Grading criteria for Special materials

To achieve a pass grade you need to:	To achieve a merit grade you also need to:	To achieve a distinction grade you also need to:
P4 identify applications of specialised materials	M4 describe the production of specialised materials	 D4 explain the implications of nanochemistry

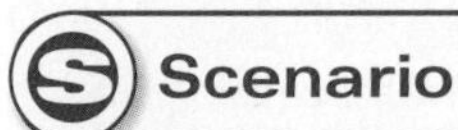

Scenario

Many building materials are being developed that use new technology. Explain to an old-fashioned builder why she should be using a range of these new materials when building a new house. Explain why the new materials are improvements on existing materials in terms of their properties. Show that modern materials can be better, cheaper and cleaner.

Figure 4.2

Keeping it local

- Go to your local sports shop and have a look at what the different pieces of equipment and specialised clothing are made from.
- Arrange an interview with someone from the medical profession. How are new materials, including nanoparticles, used in medicine?
- Find out which local opticians supply unbreakable spectacles and compare the prices with those of ordinary frames.
- Find out how new materials are used in building and fashion design locally.

Careers

- Clothing designer
- Materials specialist
- Sports therapist
- Fashion designer
- Architect
- Optician

Background science for the assignments

Traditional materials

A lot of clothing is made from cotton. Cotton is a natural material that has been used for hundreds of years.

Figure 4.3 Cotton is an example of a natural material.

Tennis rackets used to be made using wood for the frames.

Figure 4.4 Tennis rackets used to be made using natural materials like wood.

These are examples of traditional materials, which do not require sophisticated production methods.

A question of size...

Some new materials use particles that are very tiny. These are called **nanoparticles**.

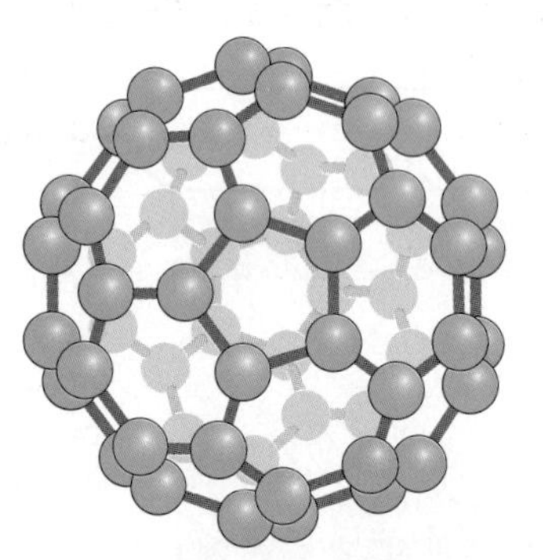

Object	Size
width of human hair	5 µm
cold virus	30 nm
buckyball	1 nm
water molecule	0.2 nm

Figure 4.5 A **buckyball** is a molecule with the formula C_{60} – it is a form of pure carbon. It is an example of a nanoparticle.

You need to be familiar with the units used to measure very tiny particles.

Measurement	Abbreviation	What it means
kilometre	km	one thousand metres
metre	m	one metre
millimetre	mm	one thousandth of a metre
micrometre	µm	one millionth of a metre
nanometre	nm	one billionth of a metre

Figure 4.6 Shape memory **alloys** can 'remember' their original shape and go back to it if bent or twisted.

Figure 4.7 Photochromic materials can change colour in different conditions.

Key words:

buckyball, micrometre, nanometre, nanoparticle

Special materials activities

1 a) Find out how the material cotton is made.
  b) What are the benefits of making clothes using cotton?
  c) What are the limitations of making clothes using cotton?

2 a) Find out what the strings and handle guard of an old-fashioned tennis racket were made from.
  b) What were the benefits of making tennis rackets from natural materials?
  c) What were the limitations of the traditional tennis rackets?

3 Work out how many metres there are in:
  a) 7 km
  b) 200 mm
  c) 1500 μm
  d) 2 000 000 nm

4 How many millimetres are there in a kilometre?

5 How many micrometres fit into a metre?

6 How many micrometres are there in a millimetre?

7 How many nanometres are there in a kilometre?

8 How many nanometres fit into a millimetre?

9 In 2003 a member of the Royal Family and some environmentalists suggested that 'self-replicating nanotechnology-sized robots' could turn the planet into a wasteland. They said this would happen because the tiny robots would break up materials into nanoparticles for making more and more tiny robots until there was just 'grey goo' left.
  a) What does 'self-replicating' mean?
  b) What does 'nanotechnology-sized' mean?
  c) Describe what you know about nanoparticles. (You may have seen the word on cosmetics or sports equipment.)
  d) The 'grey goo' idea was first suggested in a work of fiction. Do you think the events described here could come true?
  e) Do you think we should ban all nanotechnology if this possibility exists?

New materials

Essential science for P4

There are many types of new materials emerging, which have a range of different **properties** and uses.

Nano-materials

Nano-materials are made from **nanoparticles**. A nanoparticle is a particle that is between 1 nm and 100 nm in size. Because the particles are so small, they have different properties from those seen in larger pieces of the same type of material. These novel properties can make nano-materials very useful.

Some sporty examples of nano-materials include silver nanoparticles in fabrics, which make clothing odour-resistant, and silicon dioxide nanoparticles, which strengthen tennis rackets when combined with carbon fibres.

Figure 4.8 Pure gold bars, made from thousands of billions of atoms.

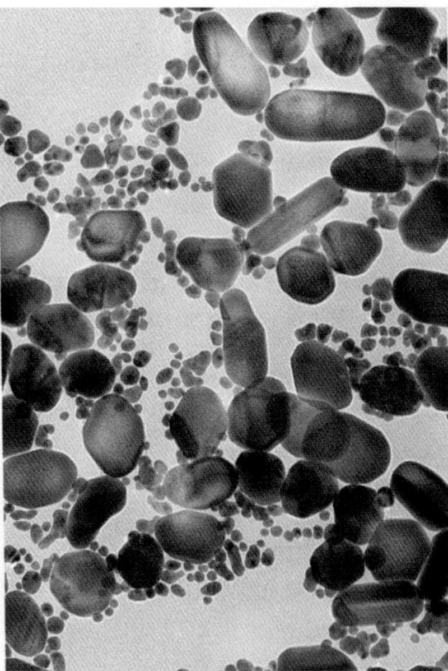

Figure 4.9 A gold nanoparticle (5 nm). It is red in colour rather than the yellow that we are familiar with.

Fullerenes are nanoparticles made entirely from carbon atoms. They can be spheres (buckyballs, see page 34) or tubes (nanotubes) and they have many varied applications.

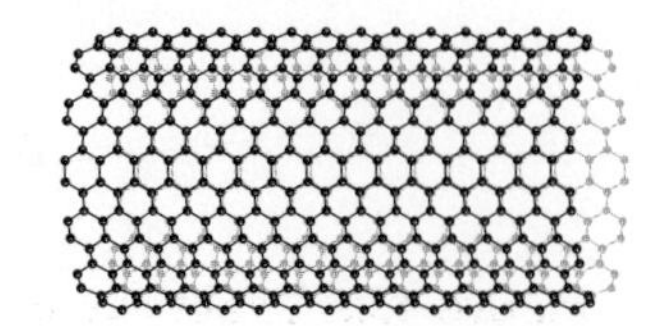

Figure 4.10 A nanotube.

Key words:

alloys, fullerenes, nanoparticles, properties

Other new materials

Figure 4.11 Kevlar® is an extremely strong material used in protective clothing.

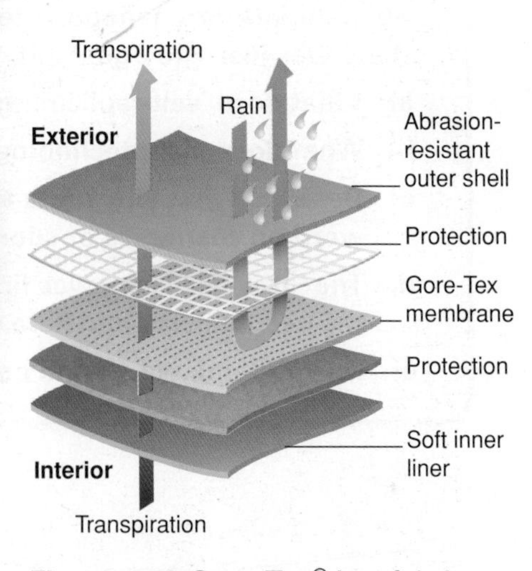

Figure 4.12 Gore-Tex® is a fabric that is both waterproof and breathable.

Task

Your task:

You need to inform a group of sportspeople about how their lives are affected, and their performances improved, by using specialised materials. You should present your information on a labelled poster.

Step 1: Research

Identify at least two products that contain nanoparticles and two products that contain other types of new materials. The products should relate to sporting equipment or clothing.

Provide a simple description of the material, its properties and why it is suitable for its sporting use. Relevant pictures and diagrams would enhance your work (but remember to reference these properly).

Step 2: Present

Design your poster:

- **Think of a catchy title, e.g. *'How sporty materials can help you!' or 'Winning materials!'***
- **Annotate pictures of sportspeople, identifying the products you have chosen and describing the material, its properties and why it is suitable for its use.**

Make the grade

P4 For P4, you must complete this task:

Identify applications of specialised materials.

Figure 4.13 These cyclists have sunglasses with lenses made of reactolite glass. This is a photochromic material that changes colour in sunlight. It is useful because when the cyclists are riding in the sun their eyes are shaded, but if it gets darker the glasses become clear so the cyclists can still see.

Making products from new materials

Essential science for M4

Sunglasses

These sunglasses have photochromic, 'reactolite' lenses. The lenses are made by embedding micro-crystals of silver chloride in the glass whilst it is molten. When exposed to UV light, the silver chloride undergoes a reversible reaction that causes it to change shape and become visible, which makes the lens darken. Originally the lenses were only available in glass, but now they can also be plastic or polycarbonate.

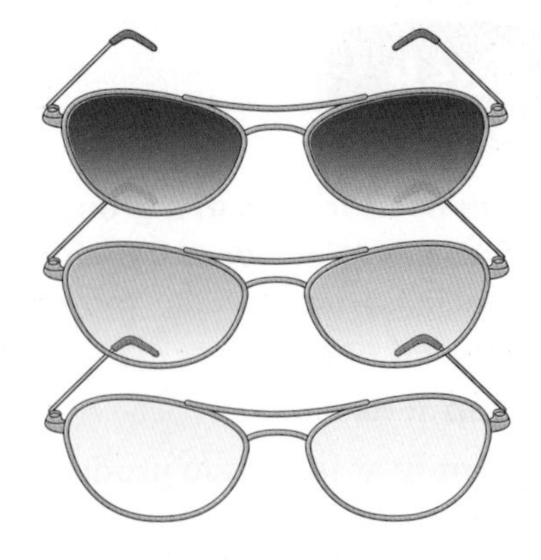

Figure 4.14 These lenses are clear in the shade, but when they are exposed to UV light (sunlight), they react and darken. The lens darkens and lightens depending on the intensity of the UV light.

Tennis racket

The frame of this tennis racket is made from carbon fibre. The gaps in the carbon fibre are filled with nanoparticles of silicon dioxide. The silicon dioxide particles stop the frame vibrating. This gives a very rigid racket that delivers much more energy to the ball.

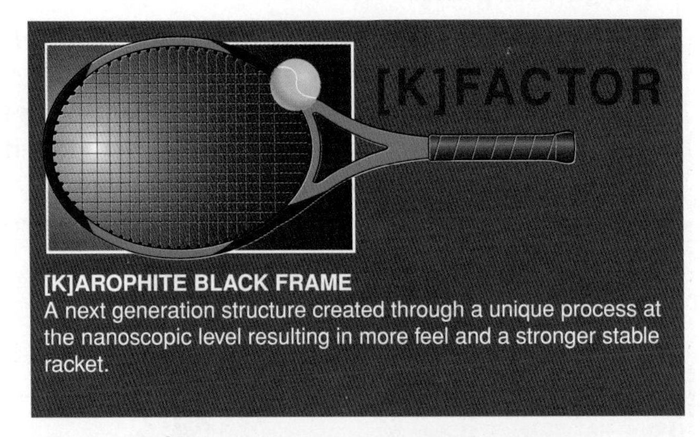

Figure 4.15 A modern tennis racket with a carbon fibre frame.

Key words:

cross-linking agent, manufacturing process, polymers

Protective headgear

This headgear contains the material Kevlar®. Kevlar® is a very hard, light material made from long strands of **polymers** that are arranged in an ordered, crystalline structure.

Figure 4.16 The outer covering of this headguard contains Kevlar® fibres.

Crystallinity is obtained by a **manufacturing process** known as spinning, which involves extruding the molten polymer solution through small holes. A **cross-linking agent** is used to promote covalent bonds between the polymer chains, linking them together to form a stronger, more rigid structure.

Task

Your task:

Make a presentation to accompany your sports materials poster. Your presentation will be given at a conference that is being attended by many different sportspeople. You will need to include a description of how your chosen specialised materials are produced and what special materials or nanoparticles they contain.

Step 1: Research

For each product containing nanoparticles, find out about the formulation of the product:

- **What nanoparticle(s) is used?**
- **What other key ingredients are there?**

For each product containing other types of new material, find out about the formulation of the product:

- **How is the new material made?**
- **What is the new material made from?**
- **Why are additives such as cross-linking agents, plasticisers or stabilisers used in the material?**

Find pictures and diagrams showing the products (but remember to reference these properly).

Step 2: Make your presentation

Design a PowerPoint presentation to accompany your speech about your chosen products.

- **Try to stick to one PowerPoint slide per product.**
- **Use the slides for pictures, diagrams and key pieces of information.**
- **Include the information about the formulation of each product in your speech.**

Make the grade

M4 For M4, you must complete this task:

Describe the production of specialised materials.

Tiny solutions or tiny problems?

Essential science for D4

A really exciting new use for nanoparticles is in pharmacy: companies are developing ways to deliver drugs directly to specific sites inside the body. This could help the drugs to last longer in the body so that doses or dose frequencies could be reduced. This would save money and could reduce the side effects associated with large doses of drugs.

However, how do we know that the nanoparticle would stay in the intended part of the body? If the particle travels to the heart or brain, it is small enough to enter cells and could do damage to the cell or to the DNA inside the cell nucleus. Currently there is little evidence about the effects on the body or the potential **toxicity** of nanoparticles.

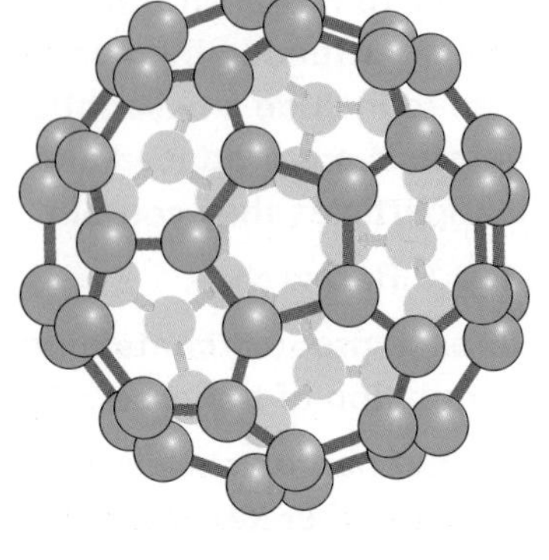

Figure 4.17 This fullerene (carbon nanoparticle) can be modified to carry drug molecules.

Figure 4.18 This scientist is working with nanoparticles.

Key words:

adverse health effects, airborne particulates, cardiovascular disease, toxicity

As the properties of nano-materials are so different to the properties of the material in bulk, there are concerns that nano-materials may have unexpected side effects. There are also huge concerns about how these tiny particles could be carried in the air and how they will affect air pollution. There is a relationship between levels of **airborne particulates** and **adverse health effects** such as asthma and **cardiovascular disease**.

Nano-materials have many new, different properties that create new opportunities, but they also present new risks and uncertainties. The increase in the production and use of nano-materials is resulting in more people being exposed to nanoparticles. Therefore, there is a greater need for information on possible health and environmental effects.

Task

Your task:

Some people are concerned about the use of new materials, in particular the use of nanoparticles. Write a newspaper report about the implications of nanotechnology. In your report you must present some of the concerns about nanoparticles.

Step 1: Research

Choose four uses of nanoparticles; two should be related to medicine. For each use, you need to find out:

The positive implications of using the nanoparticles:

- **How humans benefit from the use of the nanoparticles.**
- **How the nanoparticles increase the effectiveness of the material or procedure that they are involved in.**
- **What positive implications the use of the nanoparticles has on society, the environment and the economy.**

The negative implications of using the nanoparticles:

- **What are the safety concerns with using the nanoparticles?**
- **How could the nanoparticles damage the environment?**
- **What ethical issues may surround the use of the nanoparticles?**

Step 2: Write your report

Think of a headline and introduction for your newspaper report. Divide your report into sections:

Either:

- **Five sections: one section for each different use of nanoparticles, describing the positive and negative implications, and then one section for your conclusion.**

Or:

- **Three sections: one section for the positive implications of all four uses of nanoparticles, one section for the negative implications of all four uses of nanoparticles, and then one section for your conclusion.**

Make the grade

For D4, you must complete this task:

Explain the implications of nanochemistry.

Unit 5

Chapter 5
Motion

Theme park rides

Figure 5.1 Is it the speed, is it the turns, is it the danger that is so much fun?

Scenario

Roller coaster rides are wonderful when you are anticipating the ride, and wonderful when you think back afterwards, but during the ride they make you shriek with sheer terror.

Roller coaster rides are full of seemingly fast speeds and tight turns, rushing you close to the ground and high into the air. Working with them could be a dream job.

You are employed at a big theme park as a safety technician. It is your job to know about all the science behind motion and to use your knowledge to keep the rides, particularly the roller coaster, and the riders safe and secure. The science of moving things is called dynamics. There are many parts of this science that apply to roller coasters: potential energy gained as the roller coaster climbs to great heights; kinetic energy that increases with the speed of the roller coaster; forces (often called g-forces) caused by the turns the roller coaster goes through; frictional forces due to the contact with the rails, and more.

You will need to be able to calculate speeds and know about forces. You will need to know exactly what gravity can do and how its force affects objects around it. And you get to go on as many rides as you like, as often as you like, and to take your friends.

Grading criteria for Motion

To achieve a pass grade you need to:	To achieve a merit grade you also need to:	To achieve a distinction you also need to:
P1 carry out an investigation into an application of the uses of motion	**M1** analyse the results of the investigation into the uses of motion	**D1** evaluate the investigation into the uses of motion in our world, suggesting improvements to the real-life application

Scenario

Figure 5.2 Skaters need protective gear.

You are working as a sports and safety equipment design technician for a big sports retailer. Your job is to design cool and beautiful sportswear, which also has the built-in safety features that are needed to protect the user. You produce equipment not just for sports like rugby and football, but also for motocross, mountain biking, ice hockey, stock car racing and many others that insist the participants wear protective gear.
You need to explain how your products work to buyers from big retail stores.

Keeping it local

- Where is your nearest theme park? Contact a local theme park and find out about their procedures for employing younger people and the safety training they give.
- Go spotting the 'No boots, no hi-vis jacket, no hard hat – no job' signs on building sites near your home or college. Why do you think the sites are so insistent on these rules?
- Survey local protective clothing suppliers:
  - for motorbikes
  - for pedal bikes
  - for roller-skaters and skateboarders
  - for sports such as rugby, American football or ice hockey.

  Can you find any others?

Careers

- Road transport laboratory technician
- Seat belt and safety systems engineer
- Crash test technician
- Safety systems technician
- Theme park builder
- Car designer
- Motor vehicle technician
- Architectural technician
- Parachute technician (armed forces)

Background science for the assignments

There are words used in movement science that have very precise meanings.

Distance

This is the length from one place to another measured in metres. Sometimes centimetres or kilometres are more sensible units. Builders often measure their materials in millimetres.

Displacement

This is how far an object (such as a car) has moved from where it started out. The units are the same as for distance.

Speed

This is how fast an object (such as a car) is moving.

It is calculated as:

$$\frac{\text{displacement}}{\text{time taken}} = \text{speed (metres per second, ms}^{-1}\text{)}$$

Velocity

This is different to speed as **velocity** includes a direction. The units are the same though. It is called a vector quantity because it has a direction as well as a value.

It is calculated as:

$$\frac{\text{displacement in a certain direction}}{\text{time taken}} = \text{velocity (metres per second, ms}^{-1}\text{)}$$

Units

Metres per second can be written as m per s, or m/s, or ms^{-1}.

Metres per second per second can be written as m per s^2, or m/s^2, or ms^{-2}.

Forces are measured in **newtons** (N). You can feel a force of 1 newton; it is the downward force on a 100 g mass.

Key words:

displacement, velocity, acceleration, newtons

Acceleration

This is how fast your speed or velocity is changing.

It is calculated as:

$$\frac{\text{change in velocity}}{\text{time taken for that change}} = \text{acceleration (ms}^{-2}\text{)}$$

Figure 5.3 Acceleration requires a force, and the blocks stop the feet from slipping.

If you looked at the speedometer in a car as a traffic jam on the motorway cleared, you might see the reading increase from 10 mph to 50 mph in the space of 8 seconds as the car speeds up.

So your acceleration is 50 mph – 10 mph = 40 mph in 8 seconds.

This is $\frac{\text{40 mph}}{\text{8 seconds}}$ = 5 mph per second

The units become harder because we normally measure velocity in metres per second.

A sprinter might increase her speed from 2 ms^{-1} to 8 ms^{-1} in 4 seconds.

$$\text{So acceleration} = \frac{(8\ ms^{-1} - 2\ ms^{-1})}{4\ s} = \frac{6\ ms^{-1}}{4\ s} = \text{1.5 metres per sec per sec}$$

We write this as 1.5 ms^{-2}. It means 'an increase in speed of 1.5 metres per second for every second while she is speeding up'.

Motion activities

1 a) When you are in a car or a bus, what can you feel as it speeds up?
b) What can you feel when you are travelling at an even speed on a smooth road?
c) What can you feel as you slow down?

2 In a lift, what can you feel as it starts and stops? Why do you think you feel these sensations?

3 Look at the graph in Figure 5.4. From the information on the graph:
a) What is happening at stage A–B?
b) What is happening at stage B–C?
c) What is happening at stage C–D?
d) What is the velocity from A to B?
e) What is the velocity from E to F?

4 a) A car accelerates from 10 m/s (22.5 mph) to 30 m/s (67.0 mph) in 6 seconds. What is the acceleration?
b) The car brakes and comes fully to a stop in 12 seconds from the start of braking. What was the acceleration?

5 a) If you weigh 60 kg, what is the force you exert on the floor in newtons?
b) Would you exert the same downwards force on the Moon? Explain your answer.

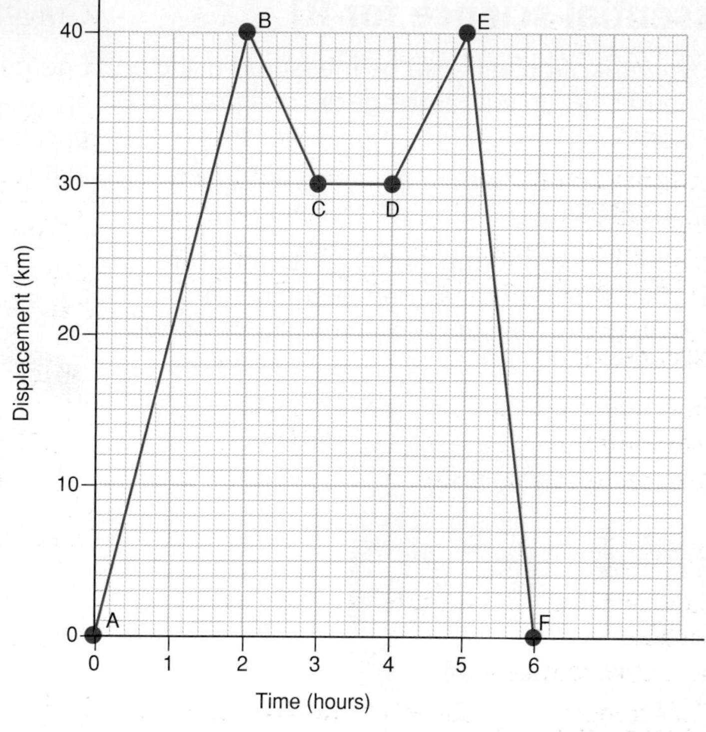

Figure 5.4 Graph of a rider's day on the Tour de France.

Figure 5.5 The speed of light is 3 hundred million metres per second. Even at this speed it takes sunlight 500 seconds to reach us – how far away is the Sun?

Measuring movement

Essential science for P1

Movement of objects can be measured using a computer set up with **light gates** or a position sensor.

Light gates

These are linked to the computer to measure how fast a card of measured length passes through them. This tells you the velocity. If two light gates are used the acceleration can be calculated from the readings.

Figure 5.6 Light gates.

Position sensor

This is linked to a computer and uses ultrasound. With it you can measure the position of a moving object up to ten thousand times a second. The sensor will allow you to plot a position-time graph using the computer. A position sensor works in a similar way to a police speed gun.

Figure 5.7 Police using a speed gun.

Key words:

gravitational potential energy, kinetic energy, light gate, friction, drag

Gravitational potential energy

Energy can be stored in an object by lifting it up against the force of gravity. This is called **gravitational potential energy**. Energy is stored in an object because of its lifted-up position.

> Potential energy = m × g × h
> m = mass of the object (kg)
> g = gravity field (10 m/s^2 on Earth)
> h = height lifted up (metres)

This is the energy that makes a roller coaster accelerate downwards when it is at the top of a slope.

Kinetic energy

A moving object has energy stored in it because it is moving.

> Kinetic energy = ½ mv^2
> m = mass of an object (kg)
> v = velocity (m/s^2)

Pendulum

A pendulum is a good example of potential energy being transferred as kinetic energy and back again.

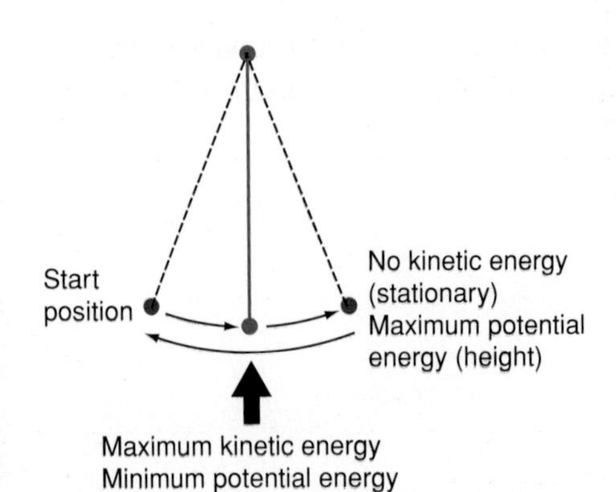

Figure 5.8

Friction and drag

Friction with the track and **drag** from the air will take some of the energy from the moving roller coaster car.

Task

Your team is designing a new roller coaster ride for a big theme park. As the roller coaster sets off it is pulled up to the top of a slope by a set of pins in the track. After that it picks up speed down a slope, pushed just by its own weight. It is kept going all the way round the track just by the speed it had at the top of the first slope.

Your task:

A big question is posed in your job as a safety technician for theme park rides. The theme park's owner, who is not a scientist, says:

- **'It's obvious that when the roller coaster is full of heavy people it will accelerate down the slopes much faster than if it was full of light people. We'll have to weigh people as they get on. Won't we?'**

Science and experience tell you the roller coaster will travel at the same speed whether it is full or empty. Can you prove this?

The owner also says:

- **'Its speed won't keep it going all the way round – it's like pushing a toy car along a carpet – it soon slows to a stop. It won't work.'**

You think the roller coaster will conserve the energy it needs to travel all the way round the ride. Can you prove this?

Step 1: Set up the apparatus

Set up a suitable practical investigation.

Below is some apparatus you could use:

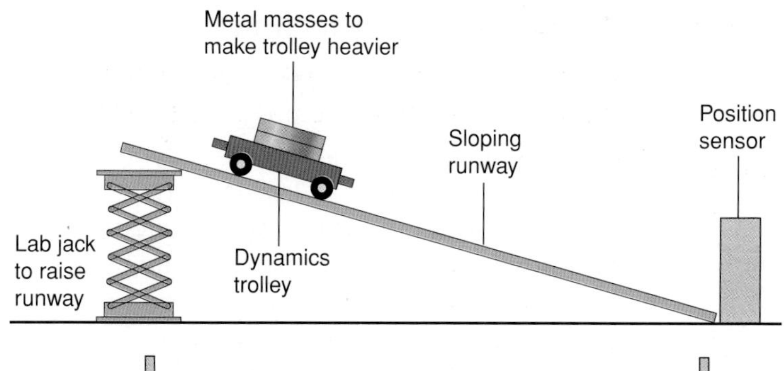

Figure 5.9 Load the trolley with different masses, and then check how fast it accelerates using the light gates or a position sensor.

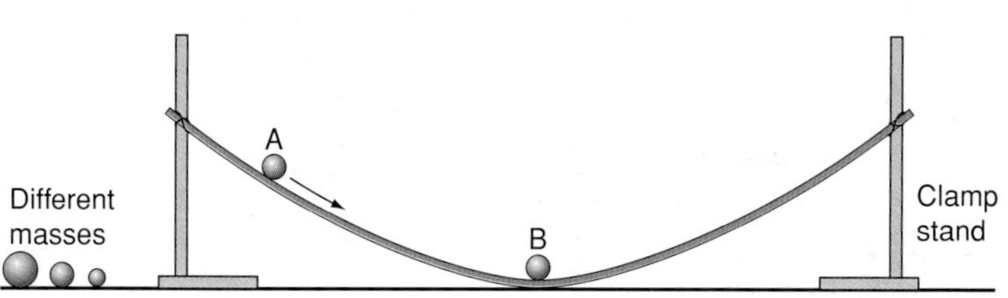

1. Use the track without ball B to see how long it takes ball A to lose energy to the surroundings.
2. Investigate the effect of energy in collisions by using balls A and B.

Figure 5.10 Use the track to check how quickly a mass will lose energy as it converts potential energy into kinetic energy and back again.

Step 2 Gather the data

Produce all the data you will need to make a presentation later. You will need to show the owner that his ideas are in fact wrong; they are misconceptions.

Make the grade

P1 For P1, you must complete this task:

Carry out an investigation into an application of the uses of motion.

Conservation of energy

Essential science for M1

Key words:
conservation of energy, pendulum, diluted gravity

When dealing with moving objects in a real situation, remember two principles:

1. **Energy is conserved** as it is transferred from kinetic energy to potential energy and back again. You can set up a simple **pendulum** experiment to show this.
2. In any energy change some energy is transferred to the surroundings as low-level heat energy. No transfer can be 100% efficient. In the roller coaster, friction and drag take some of the energy and transfer it to the surroundings as low-level heat and sound energy.

Graphs

A good visual way of analysing movement data from your investigations is to plot graphs of the data.

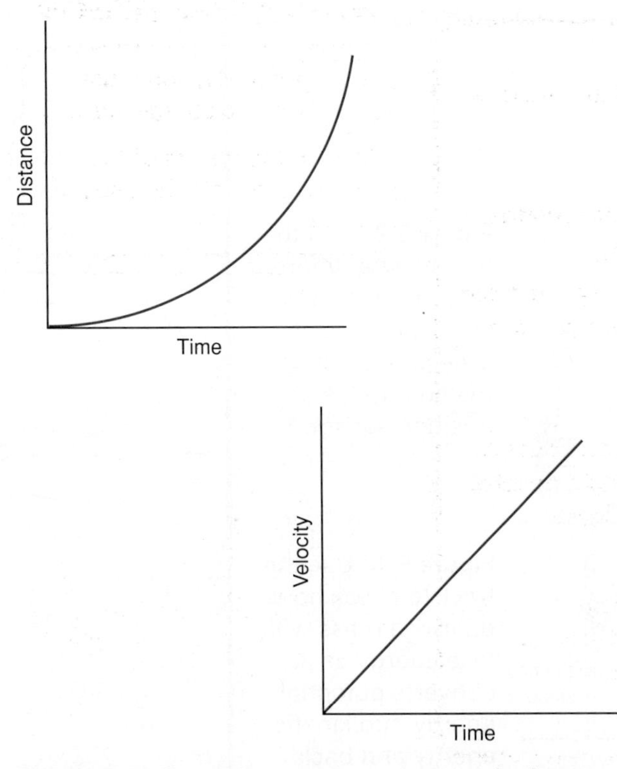

Figure 5.11 From the graphs you can calculate slopes and use this data to calculate velocities and accelerations. Note: both graphs are of the same accelerating object.

Diluted gravity

The force making a trolley roll down a slope can be thought of as 'diluted gravity'. The trolley, or roller coaster car, moves to the bottom of the slope because gravity is pulling it downwards. The trolley or car moves along horizontally because the runway is in the way and it has to 'go along' to 'go downwards'.

The steeper the slope, the less diluted the gravity.

Compensating for friction

You can't get away from friction and drag: their effect is with us always. It steals the kinetic energy that costs money to create and transfers it to the surroundings as low-level heating.

You can set up your trolley runway to compensate for the effects of friction. Give your runway a very small slope. Then give the trolley a gentle push. If you have exactly balanced the effect of friction with diluted gravity, the trolley will move all the way along the runway without speeding up.

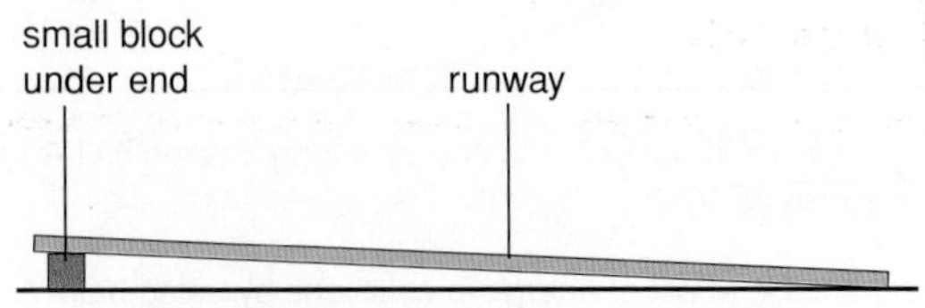

A very slight slope will compensate for friction. A trolley should run down the slope when pushed without stopping due to friction or accelerating due to gravity.

Figure 5.12 Compensated runway.

Friction and drag

Engineers constantly battle against friction to reduce energy losses. Of course, when you want to stop an object, friction is the engineer's friend.

Friction happens between objects and the solid surface they are moving across. Drag happens for objects moving through a gas or liquid.

Friction and drag are major problems for dynamics investigations. They can make the results very difficult to draw valid conclusions from.

One method is to reduce friction and drag to a minimum.

- Find out about air tracks and moving objects supported by a cushion of air.
- Think about moving objects on ice, where the friction is greatly reduced.
- Think about objects like satellites moving in space above the Earth's atmosphere.

Task

Your task:

In addition to collecting data for your P1 investigation, you now need to analyse and present that data.

Step 1: Results

Draw graphs of the data you have gathered or present graphs drawn for you by the computer used for data logging.

Provide a script or a panel of writing to explain exactly what the graphs are showing and how this relates to the views you are trying to disprove.

Step 2 Conclusions

Write a conclusion for each one of your investigations. Explain fully what your results show and why your data is valid despite the influence of drag and friction on the moving objects.

Step 3 Present

Make a visual presentation, explaining to the owner that the ideas he has are in fact misconceptions. Use the graphs you have produced from your data and any other visual presentation aids.

Make the grade

M1 For M1, you must complete this task:

Analyse the results of the investigation into the uses of motion.

Did I do well?

Essential science for D1

Evaluating what you have done in your motion investigations means you must look closely at what you have produced. Then you must justify the methods and equipment you chose to use.

In this section you will need to evaluate your investigations into the application of movement science and dynamics, and try to suggest how to improve some real-life applications.

Use the following questions to help you plan what you will write.

Questions

1 What was my purpose in these investigations?
2 What scientific knowledge did I use to plan the work?
3 What scientific skills did I use to carry out the investigations?
4 Were my investigations successful? Did I get good **data**?
5 Are my **conclusions** valid?
6 Would I change any part of my investigations, and if so which part(s)?
7 How would I improve the investigations if I had time to do them again?
8 What degree of confidence have I got in the results of my investigations? (Think about: If I say it's OK, people including my friends will be hurtled about in theme park rides at high speed and with tight turns. If I'm wrong, people may be injured or killed.)

Figure 5.13 Knowing about the science of movement will make speeding up, travelling and stopping much safer. One thing that repeatable science data gives you is the ability to make **predictions**.

Key words:

investigate, predict, data, conclusion

Task

In addition to the investigations you carried out for P1 and the analysis you did for M1, you now have to evaluate both of your investigations and suggest some improvements to the real-life applications.

Your task:

Step 1: Evaluate your investigations

Look closely at the results (P1) and analysis (M1) for both of your investigations.

Use the questions on page 50 to evaluate both of your investigations.

Step 2: Conclude

Make an overall evaluation of your investigations. Decide if the movements in your investigations were uniform or non-uniform motion.

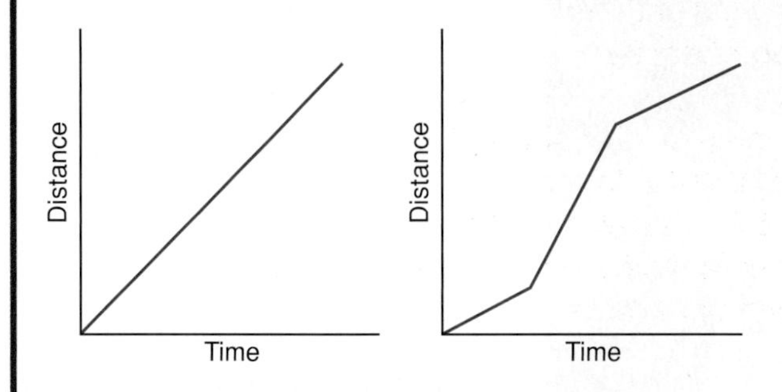

Figure 5.14 The difference between uniform and non-uniform motion.

Step 3: Present

Write up your evaluation. You may wish to add this section to the reports you wrote for P1 and M1.

Step 4: Real-life applications

Here are some 'real' applications that you can apply your conclusions to:

- **stopping distances of motor cars**
- **high speed train lines**
- **racing car and sports car technology**
- **track and tour de france cycle racing**
- **speeding up submarines**
- **parachutes and drag chutes on fast moving vehicles like the space shuttle**
- **safety systems in vehicles such as trains**
- **making buildings to withstand high winds**
- **and of course – theme park rides.**

When you have finished, show your teacher.

Make the grade

D1 For D1, you must complete this task:

Evaluate the investigation into the uses of motion in our world, suggesting improvements to the real-life application.

Unit 5

Chapter 6
Forces

Crash test dummy

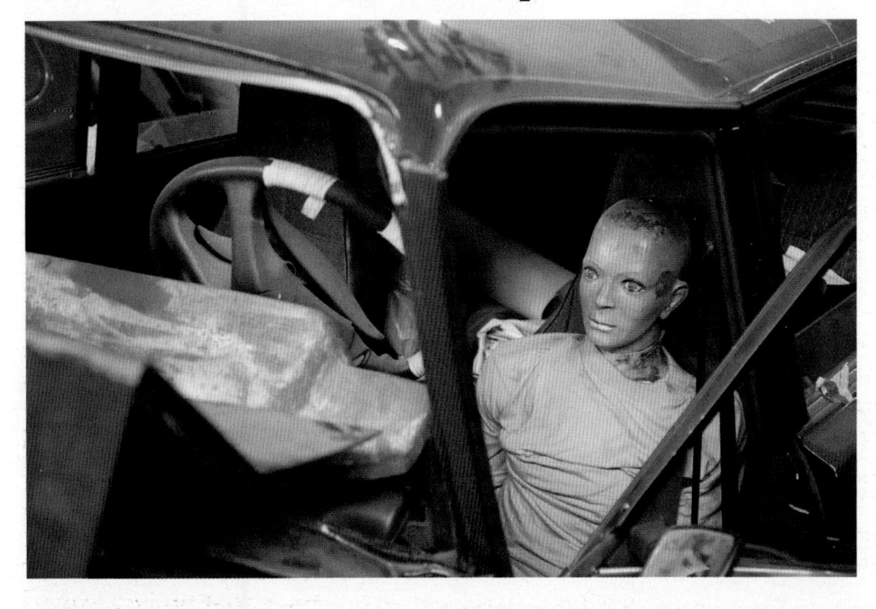

Figure 6.1 Here is someone who has a lot of experience of the effects of forces.

Scenario

Figure 6.1 shows a crash test dummy. You often see him on TV when he is hard at work, and hard is often an appropriate word. He gets banged, bashed, dropped, stretched, sent up in rockets, crushed and generally treated quite badly. But he sees it as part of a day's work and never once complains; there may even be a slight smile on his face.

You are a technician. You are going to be working as crash test dummy's minder and operator at a research facility like the Road Research laboratory. But this one is concerned with much more general risky situations, not just car crashes. There are many risky situations for people in twenty-first century life. High speed travel, tall buildings, domestic and industrial accidents, and dangerous sports all give rise to a large number of situations where people are at risk from forces acting on them. Human beings are normally quite robust animals. But flesh can squash and bones can break all too easily if you meet forces in the wrong way or without diluting their effect.

Finding out how to cope with the forces we meet in everyday life is the job for you and your crash test dummy.

Grading criteria for Forces:

To achieve a pass grade you need to:	To achieve a merit grade you also need to:	To achieve a distinction you need to:
P2 carry out an investigation into an application of the uses of force	M2 analyse the results of the investigation into the uses of force	D2 evaluate the investigation into the uses of force in our world, suggesting improvements to the real-life application

Scenario

You are part of the Forensic Science Service specialising in road accidents and other mishaps that have injured people. You are employed by insurance investigators to assess the damage to people and property caused by accidents. To carry out your job you need to know a great deal about the way forces are created by accidents – in any collision it is only the last five centimetres that hurt! In addition, you need to know a lot about the properties of materials and the effects of forces in order to analyse the damage done.

Figure 6.2 Safety headgear. Your skull is quite hard, but just inside it is a brain that bruises very easily. The job of safety headgear is to absorb and dilute the impact, often cracking as it does so to absorb some of the force. **Never** reuse a helmet after an accident. It will be weakened.

Keeping it local

- What safety services are available in your local area? Look up 'safety equipment' at www.yell.com.
- Contact your local council's Environmental Health Department to find out what services they provide.
- Contact a local policeman to find out what safety precautions (like motorcycle crash helmets) are required by law.

Careers

- Crash test technician
- Safety systems technician
- Paramedic
- Orthopaedic health worker
- Fracture clinic technician
- Physiotherapist
- Cycle or motorcycle retailer
- Parachute technician (armed forces)
- Road transport laboratory technician
- Seat belt and safety systems engineer
- Health and safety officers (local council)
- Road traffic accident investigators (Forensic Science Service)

Background science for the assignments

Figure 6.3 No matter how strong a rope is, it will break with a big enough force. No matter how strong a steel bar is, it will bend under a big enough force.

Types of forces

A force is a push or a pull. That's simple, but what a push or a pull can do depends on what is being pushed or pulled. There are different types of forces:

- **Compressive** forces squeeze objects.
- **Tensile** forces stretch objects.
- Balanced forces hold objects still.
- Unbalanced forces make objects move.
- **Moments** or torque forces make things turn.
- **Stress** is a force that changes an object's shape.
- **Friction** is a force between two solids moving past each other.

What can forces do?

Forces can do these things to an object:

- change the velocity or start it moving
- change the direction of movement
- make it bigger or smaller, longer or shorter
- change its shape
- make it turn or spin.

What is a moment?

Moment is a physics word that can cause problems. When we use the word in a forces context, it's not about 'time', but about the turning effect of a force.

Key words:

compressive, tensile, moment, stress, friction

To have a turning effect, one part of an object has to be a pivot, like the hinge of a door or the heel of a crowbar or the axle of a wheel. The pivot is fixed and the object turns round it.

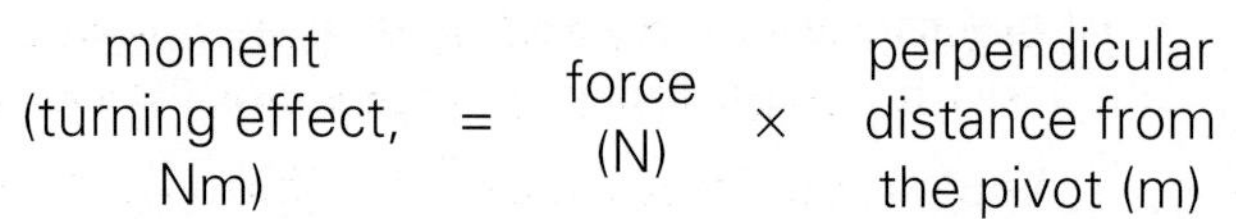

$$\text{moment (turning effect, Nm)} = \text{force (N)} \times \text{perpendicular distance from the pivot (m)}$$

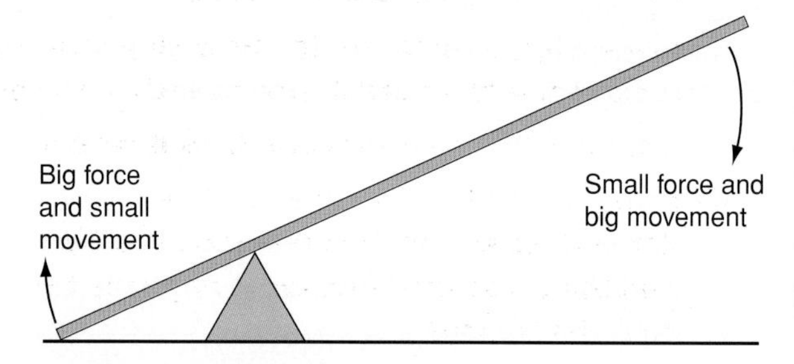

Figure 6.4 Turning effect.

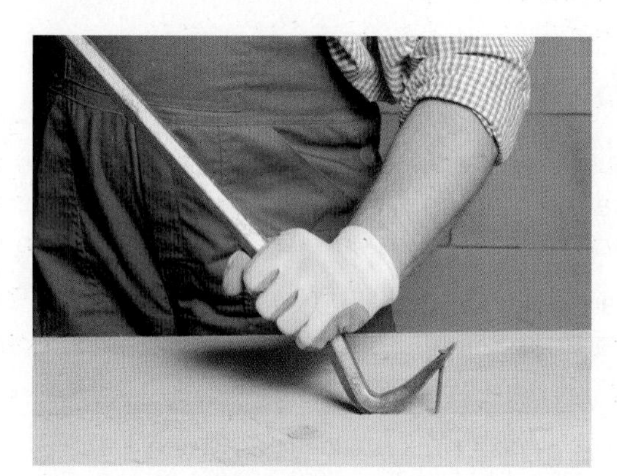

Figure 6.5 You couldn't pull the nail out with your hand. But the crowbar multiplies the force you exert by using moments. Many tools use moments to produce a big force – pliers, screwdrivers, scissors, tin openers and more.

Forces activities

1 Elastic materials change shape when pushed or pulled, then go back to their original shape when the force is taken away. 'Plastic' means the opposite of elastic. Plastic materials change shape, but stay in their new shape when the force is removed. An example of this is clay.

a) Name two elastic materials, and two materials other than clay that have the opposite 'plastic' property (not plastic as in a polymer).

b) What forces can you feel as you stretch a rubber band, at first and then near its breaking point?

c) Do you feel the same forces when stretching a spring?

d) If you bend a plastic ruler, describe its properties up to the point when it snaps.

e) Do you get the same properties if you bend a metal ruler? (Try this out on a strip of metal that is not a ruler if necessary!)

f) Experiment with bending dry, uncooked spaghetti. Form a theory about why it always snaps into three pieces.

Figure 6.6 We use the force from elastic bands in lots of applications.

2 a) Look at the forces shown in Figure 6.7. What are the sizes of the forces (A and B) involved?

b) Draw a forces diagram to show someone pushing a sofa across a carpeted room. Draw the forces for just before it moves and as it is beginning to move.

c) Explain what friction is and how it works.

d) If you push a four-wheeled trolley on a flat surface so that it is moving steadily and then stop pushing, what happens to stop the trolley?

e) You use a small booster to push a spacecraft in space so that it is moving steadily, and then stop the booster. Explain what happens to the movement of the spacecraft.

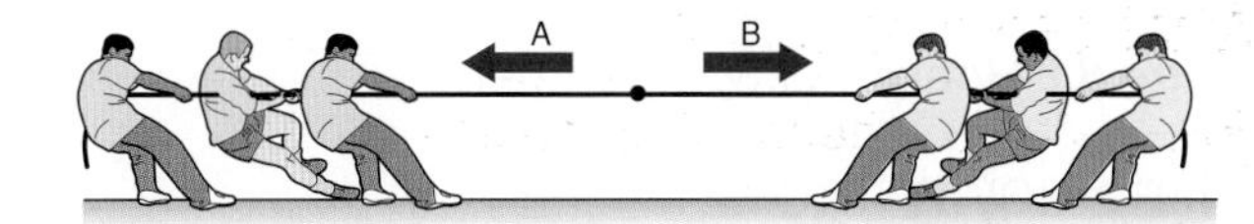

Figure 6.7 Forces in a tug of war match.

3 You have probably experienced trying to run in water, so you know how difficult it is. Even air is 'thicker' than you might think. There is a kilogram of air in the space under your table or desk, and a school hall contains about 6 tonnes of air.

Think about friction and drag and answer these questions.

a) Why do you oil the hubs of your bike wheels?

b) What shape are fish, dolphins and penguins? How does this help them to move through water?

c) What shape are fast cars and aeroplanes?

d) Why do aeroplanes fly so high above the ground?

Forces investigations

Essential science for P2

Stretching forces

Use the apparatus shown in Figure 6.9 to investigate how much certain materials change shape when a force is applied to them. Gather data and take measurements to show how much a spring, wire or elastic band will change in length.

Investigate the **elastic limit** of materials, and investigate what happens to the properties of materials near their elastic limit.

Investigate how copper wires and nylon filaments behave near their breaking points.

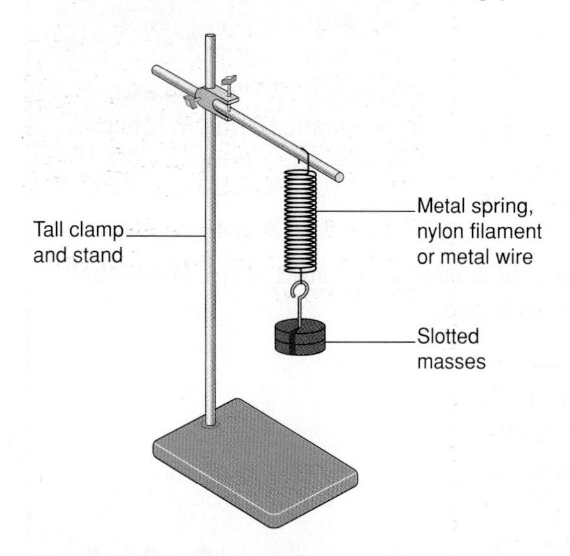

Figure 6.8 Hooke's Law experiment.

Falling down forces

The situation when falling through the air is different. You are unlikely to land on a soft, crumply surface unless you are a pole vaulter. But air is thicker than you would think. Investigate using the drag from moving through the air to slow down a falling crash test dummy.

Key words:

stretch, elastic limit, acceleration, deceleration, crumple zone

Acceleration forces

Acceleration happens according to a law of physics, which states that:

- bigger pushes accelerate objects faster
- heavier objects are harder to accelerate.

This law can be shown in an equation:

force (N) = mass (kg) × acceleration (ms^{-2})

The equation still works if you are looking at the forces needed to **decelerate** a fast-moving object, such as a crash test dummy's head during a crash.

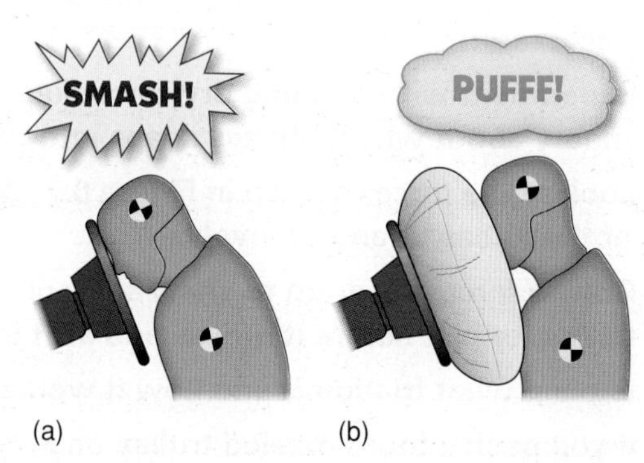

Figure 6.9 Forces on a crash test dummy's head during a crash (a) with no air bag and (b) with an air bag.

In Figure 6.10(a) the deceleration force =

$$10 \text{ kg} \times \frac{30 \text{ ms}^{-1}}{0.01 \text{ s}} = 30\,000 \text{ N}$$

(equivalent to a weight of 3 tonnes).

In Figure 6.10(b) the deceleration force =

$$10 \text{ kg} \times \frac{30 \text{ ms}^{-1}}{0.05 \text{ s}} = 600 \text{ N}$$

(equivalent to a weight of 60 kg).

You can see that in a crash, it is not necessarily the velocity but how quickly you slow down that is the danger factor.

Collect data about **crumple zones** in cars.

Investigate packaging an egg so it will not break when dropped from a height of 1 metre onto a hard floor.

Task

You will carry out an investigation into one (or more) of the following applications of forces:

- **stretching materials**
- **crumple zones**
- **falling down.**

Your task:

Step 1: Planning

You are going to measure the effects of forces in some of the investigations shown on page 56.

Decide which investigation(s) you are going to carry out. You may have time for more than one. It does not need to be one of the investigations shown here; you could adapt the methods to a real-life situation (e.g. bungee jumping).

Plan your investigation(s) carefully. Carry out trial runs to find out the limits of your apparatus. Carefully write down exactly what procedures you are going to use and what you intend to measure. It is very important that other researchers will be able to repeat your measurements later. Repeatable data has a high degree of validity when it comes to persuading someone that your conclusions are valid.

Step 2: Carrying out the investigation(s)

Carry out the investigation(s) carefully, repeating measurements where you can. You must record exact details of what you do and make a careful note of any measurements you take. An investigator would use a digital camera or video phone camera to record precise images for later use.

Remember, these investigations may be used to help people survive very hazardous situations in fast cars, dangerous sports and accidents.

Step 3: Saving your data

Store your data and any images carefully; they are a valuable resource. You will need them in the later tasks.

Make the grade

P2 For P2, you must complete this task:

Carry out an investigation into an application of the uses of force.

Figure 6.10 A pole vaulter uses several types of force. Can you decide how?

Working with forces

Essential science for M2

Vector quantities

Forces have both size and direction – and both are important. Force is a **vector quantity**. For example, in Figure 6.7 on page 55, if the two teams were both pulling in the same direction the **resultant** force would be very large. As it is, the overall force is zero and the rope does not move as the two forces cancel each other out.

If a sailing ship is pulled by a tug boat in an easterly direction and at the same time pushed south by a strong wind, what direction does it travel in?

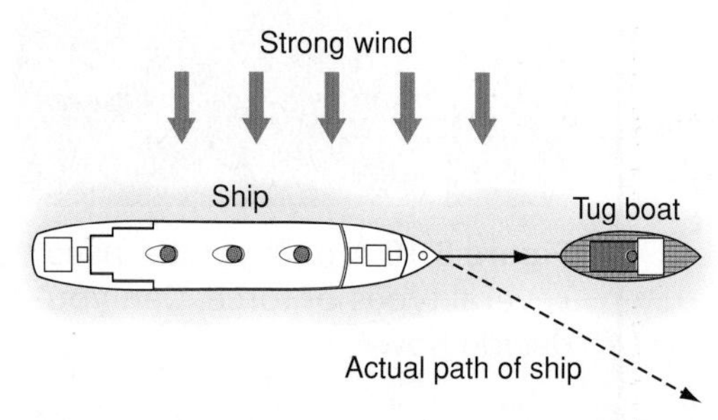

Figure 6.11 To work out the direction of the resultant force takes a scale drawing or complicated calculation.

Tip
Resultant force = effect of adding two or more forces together. You may need a scale diagram to find out the resultant as the directions of the forces may not be the same.

Key words:

vector quantity, resultant, random errors, systematic errors

Analysing the data

To analyse the data from your investigations you will need to use both calculations and graphs. For example, to justify any conclusion in the crumple zones investigation, you will need to show calculations like the forces calculation on page 56. This will show the effect on the deceleration force of spreading out the 'slowing down' over a longer period, even if that period is fractions of seconds. For the falling down investigation you will need to measure and calculate the speed at which your parachute or other device falls through the air.

Figure 6.12 Crumple zones and air bags save lives.

Graphs

No experimental data is perfect and there will always be errors in the measurements. Some of these errors are **random**, due to human error and chance factors. Other errors are **systematic** – because of the way the experiment is designed.

When you put your data points on a graph, remember that the movements you saw of stretching, crumpling and falling objects were usually smooth rather than jerky movements. So make your graphs smooth lines through the points, never jerky or join-the-dots. Smoothing out the lines in your graphs reduces the effect of random errors. To reduce systematic errors you need to carefully look at the experimental methods.

Task

In addition to collecting data for your P2 investigation(s), you now need to analyse and present that data.

Your task:

Step 1: Produce graphs and tables

Draw graphs of the data for any investigation(s) you have carried out. Produce tables of any calculations you have made. Graphs are best because our visual minds see patterns in pictures better than in blocks of numbers.

Look carefully at the graphs and try to decide the story they tell of how the variables relate to each other.

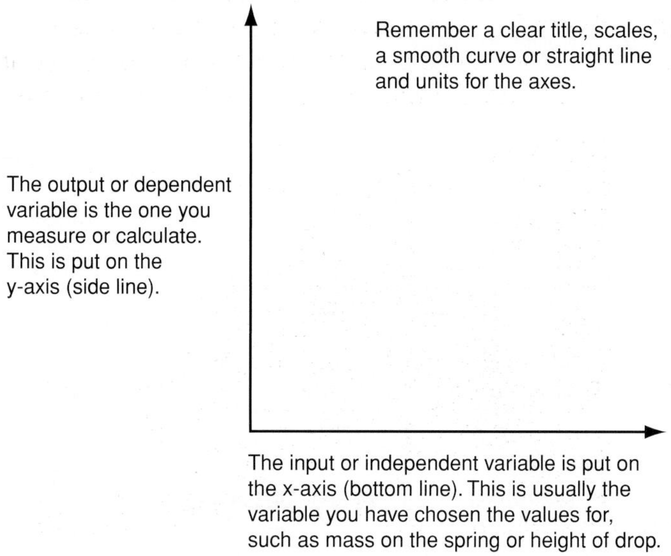

Remember a clear title, scales, a smooth curve or straight line and units for the axes.

The output or dependent variable is the one you measure or calculate. This is put on the y-axis (side line).

The input or independent variable is put on the x-axis (bottom line). This is usually the variable you have chosen the values for, such as mass on the spring or height of drop.

Figure 6.13 Points to remember when drawing your graphs.

Step 2: Write conclusions

Write a clear conclusion for each investigation. You will need to be able to justify your conclusions and explain why the evidence you are using is valid and how it has been gathered carefully.

Step 3: Present

Make a visual presentation for each of the investigations. Include any images or video you may have captured while doing the experiments.

Make the grade

For M2, you must complete this task:

Analyse the results of the investigation into the uses of force.

Did I do well?

Essential science for D2

In this section you will need to evaluate your investigations into the application of forces, and try to suggest how to improve some real-life applications.

To evaluate what you have done in your forces investigations, you must look closely at what you have produced. Decide if what you have found out makes sense and whether your conclusions are sensible and fit in with what you feel. Then you should justify the methods and equipment you chose to use.

Use the following questions to help you plan what you will write.

Tip
Remember: unlike many other scientific phenomena, your body can 'feel' forces. Your conclusions and results should fit in with what you can feel.

Questions

1. What was my purpose in doing these investigations?
2. What scientific knowledge did I use to plan the work?
3. What scientific skills did I use to carry out the investigations?
4. Were my investigations successful? Did I get good data?
5. Are my conclusions valid?
6. Would I change any part of my investigations, and if so which part(s)?
7. How would I improve the investigations if I had time to do them again?
8. What degree of confidence have I got in the results of my investigations? (Think about: if I say it's OK, people including my friends may rely on the safety advice I give as a result of the investigations. If I'm wrong, people may be injured or killed.)

Key words:

air pressure, data, skills, conclusion, validity, evaluation

More science

Think carefully about how the ever-present force of gravity affects your investigations. It will make a difference to all your investigations. For example, frictional forces are due to gravity squashing surfaces together. In zero gravity there is little friction. The Earth's gravity field is 10 N/kg at the Earth's surface.

Also think carefully about air pressure. Air is quite heavy and is pushing down on us.

Our atmosphere exerts a force on all of us of 100 000 N/m^2 (that's 10 tonnes of weight per square metre!). This makes the atmosphere thick and is responsible for the drag forces we cannot avoid in our investigations.

Figure 6.14 Moving air can cause forces that damage strong objects.

Task

In addition to the investigations you carried out for P2 and the analysis you did for M2, you now have to evaluate both of your investigations and suggest some improvements to the real-life applications.

Your task:

Step 1: Evaluate your investigations

Look closely at the results (P2) and analysis (M2) for both of your investigations.

Use the questions on page 60 to evaluate both of your investigations.

Step 2: Conclude

Make an overall evaluation of your investigations. What were the conclusions about how you can manage the effects of forces in modern life to make people and places safer?

Step 3: Present

Write up your evaluation. You may wish to add this section to the reports you wrote for P2 and M2.

Step 4: Real life

Some 'real' questions that you can apply your evidence to include:

- **Should motor cars be made like bumper cars?**
- **Should the top speed limit be reduced to 30 mph?**
- **Can you suggest easy ways to escape from tall buildings?**
- **Car crashes are unavoidable. How can damage to people be prevented?**
- **What safety equipment can be used in sports like ice hockey, motorcycle racing or bicycle road racing to prevent or reduce injuries?**
- **Is it true that if crash helmets were big and soft they would be more effective? Would people wear them?**
- **Is risk of injury part of the enjoyment of some activities?**

When you have finished, show your teacher.

Figure 6.15 Is the risk of free climbing part of the enjoyment?

Make the grade

For D2, you must complete this task:

Evaluate the investigation into the uses of force in our world, suggesting improvements to the real-life application.

Unit 5

Chapter 7
Light and sound

Bright lights and loud sounds

Figure 7.1 Without the sounds and lights this would just be a big room.

Light and sound are vital to so many aspects of our lives. They help us to stay safe, find our way around, communicate in many different ways, learn about the Universe and enjoy ourselves.

Light and sound are forms of wave energy. Light waves are a form of energy we can see and sound waves are a form of energy we can hear.

At a nightclub, light and sound combine together. Disco balls reflect light, and sound travels from many different loudspeakers.

Scenario

You are a scientific research assistant who works for a 'light and sound' engineering company. This is the sort of company that does the stage set-up for big shows at the O2 Arena for Lady Gaga. There are a lot of different electrical and building technicians who work with you.

Part of your role is to produce information for the company brochure and show some of the ways in which light and sound waves are used in everyday life.

Your reports should include pictures, diagrams, and also the results from scientific investigations into the many uses of light and sound.

Grading criteria for Light and sound

To achieve a pass grade you need to:	To achieve a merit grade you also need to:	To achieve a distinction you also need to:
P3 carry out an investigation into an application of the uses of waves	**M3** analyse the results of the investigation into the uses of waves	**D3** evaluate the investigation into the uses of waves in our world, suggesting improvements to the real-life application

Scenario

You are an optical lens technician working for an optical lens company. Part of your role is to find new materials that can be used to make the lenses for both glasses and contact lenses. You will need to carry out investigations on concave and convex lenses in order to test the new materials.

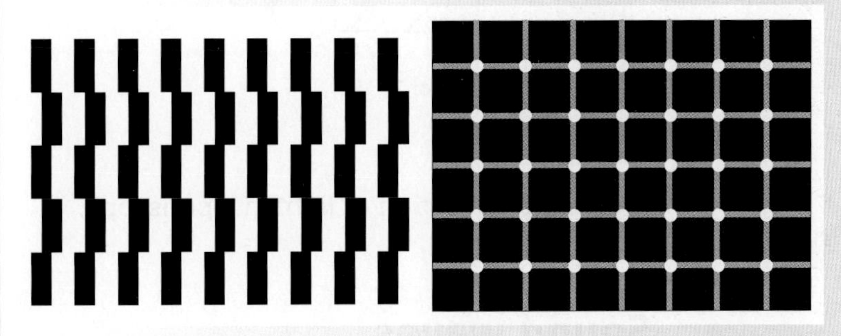

Figure 7.2 Optical illusions like these can be fun but also quite confusing. This is because our brains are trying to make sense of what we are seeing. Do the lines across the image on the left look straight? Now use a ruler to check if they are. Can you count the black dots in the grid on the right?

Keeping it local

- Look up how many opticians there are in local shopping centres. Are there a similar number of services helping people with hearing problems? Investigate why you think this is?
- Look up what local companies supply sound systems or lighting systems for leisure locations like restaurants, sports stadiums and music venues. What skills do people need to work for these companies?

Careers

- Optician
- Lighting technician for festivals
- Sound system engineer for festivals
- Radio or TV broadcast technician
- Music venue designer
- Medical technician – eye and ear problems

Background science for the assignments

Light waves

Light is a form of **energy** and is part of the **electromagnetic** (EM) **spectrum**. Light is the visible part of the EM spectrum, and our main light source on Earth is the Sun. Without the light from the Sun, there would be no life on Earth. Light does not need a **medium** and can travel through a **vacuum**.

Applications of light waves

Reflection

Mirrors, periscopes and kaleidoscopes use reflection.

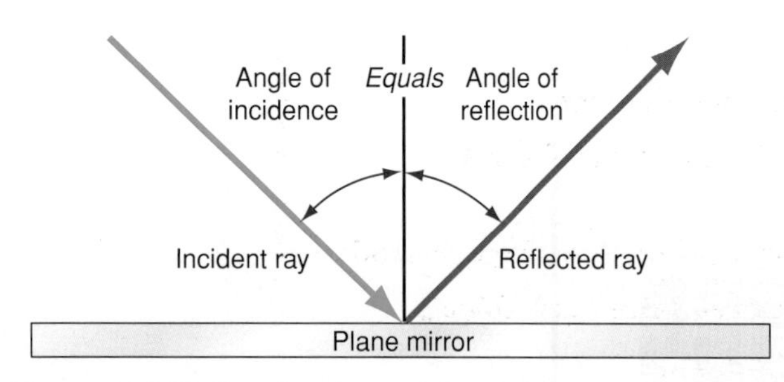

Figure 7.3 Reflection in a plane mirror.

Internal reflection

Optical fibres and cat's eyes use internal reflection.

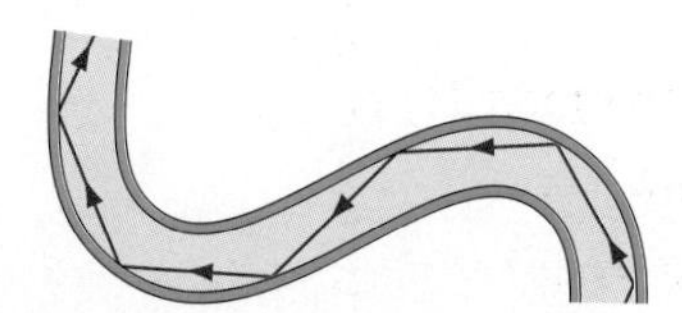

Figure 7.4 Total internal reflection along an optic fibre.

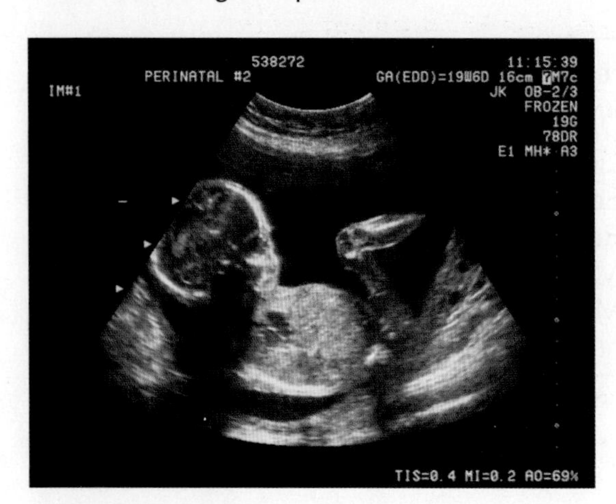

Figure 7.5 Ultrasound can be used to see a developing baby.

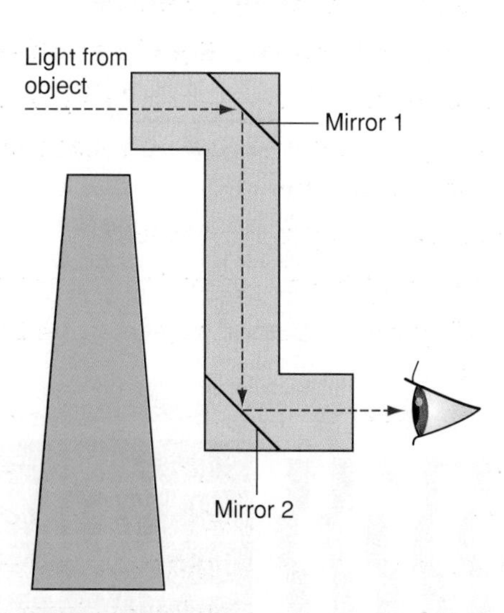

Figure 7.6 Reflection of light in a periscope.

Sound waves

Like light, **sound** is a form of energy. Sound travels as waves of **vibrations** through the molecules that make up substances like air. Sound cannot travel in a vacuum – because there are no molecules to pass the vibrations on.

Applications of sound waves

Low-power **ultrasound** beams can be beamed into the body and a computer can analyse the echoes to produce an echo-cardiogram image or 'scan' of an unborn baby or of organs like the heart. High-power ultrasound is used to break down kidney or bladder stones in the body without the need for surgery.

A **sonar** system sends out pulses of sound waves and detects returning echoes. By knowing the speed of sound in different materials such as air and water, scientists can work out how far away objects are. Animals such as bats use **echolocation**.

Light and sound activities

1 Carry out some of these practical experiments using mirrors, different glass blocks and a ray box.
   a) Reflection of light in a plane mirror.
   b) Refraction of light through a glass block.
   c) Refraction of light in water.
   d) Dispersion of light through a prism.
   e) Total internal reflection of light in a right-angled prism.

2 a) Name the seven colours that make up visible light.
   b) Describe a practical investigation that can be used to split white light up into these seven colours.

3 What is the speed of light in a vacuum?

4 Try to find out what the speed of light is in water and in glass.

5 Look at the diagram of a periscope on page 64. Try to explain how a periscope works.

6 Match these types of light production to the way that the light is produced:

incandescence	light released by glow-in-the-dark objects
fluorescence	light produced by animals
phosphorescence	light produced by hot objects like the Sun
bioluminescence	giving out light without getting hot

7 Name three objects that rely on the reflection of light to make them work.

8 Explain (using a diagram) why an object at the bottom of a swimming pool appears to be closer to the surface of the water than it really is.

9 What is the name for the way that light appears to bend as it enters water?

10 Find out how telescopes work. How do they use light to make them work?

11 What is the normal range of sounds that humans can hear? Choose the correct answer.

   20 Hz–20 000 Hz   5000 Hz–50 000 Hz   900 Hz–9000 Hz

12 Dog trainers often use an ultrasonic whistle that can be heard by dogs but not by humans. Why can dogs hear the sound but humans can't?

13 Ships and submarines use sonar (echolocation) to find out the depth of water. How else can sonar be useful to ships and submarines?

14 Why is ultrasound a really useful tool in the medical world? Give examples.

15 What might the frequency of an ultrasound be? Remember, it has to be a high frequency that the human ear cannot detect.

16 Why does sound travel faster in solids than in liquids, and faster in liquids than in gases?

Key words:

echolocation, electromagnetic spectrum, energy, light, medium, sonar, sound, ultrasound, vacuum, vibration

Applications of light and sound

Essential science for P3

The human eye

The normal human eye works to focus or 'refract' light emitted from objects onto the **retina**. The light rays first pass through the clear front 'windshield' of the eye, called the **cornea**. They then pass through the **pupil**, which is an opening created by the **iris** (the coloured part of the eye). Next, the light passes through the **lens** and finally arrives at the **retina**. When the light rays are clearly **focused** on the retina the **optic nerve** can send a message to the brain and the result is a clear visual image. Not everyone has perfect vision, which is known as 20/20 vision.

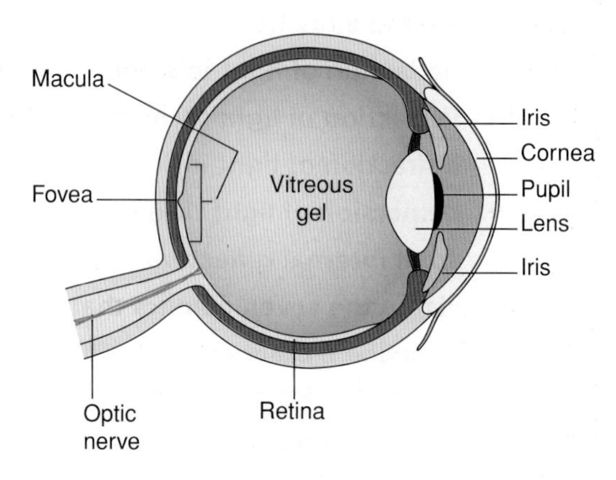

Figure 7.7 The human eye.

Using lenses to correct vision

If the image is not clearly formed on the retina, it can cause a blurred image. Lenses are used to correct the vision of both short-sighted and long-sighted people.

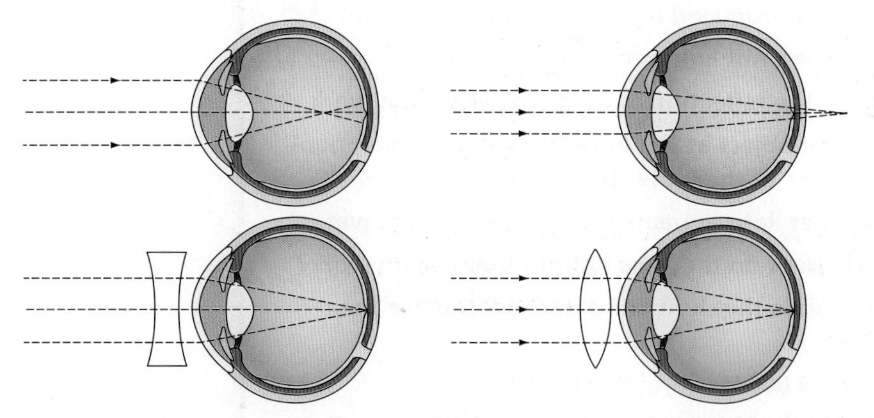

Figure 7.8 Concave and **convex** lenses are used to correct vision by causing light rays to **diverge** or **converge** onto the retina. Concave lenses correct short sight. Convex lenses correct long sight.

Key words:

cornea, iris, lens, optic nerve, pupil, retina, concave, diverge, convex, converge, focus, telescope, vibration, wave energy, vacuum

Sound

Sound is also a form of **wave energy**. Sound travels as waves of **vibration** through atoms and particles. As each atom vibrates it bumps into another, transferring its energy to them. This is why sound travels faster in solids than in gases, because the particles are closer together in solids. It is also the reason why sound cannot travel in a **vacuum**, where there are no particles.

You can use a bell jar attached to a vacuum pump to investigate the need for a medium. When the air has been removed from the bell jar, can you hear the bell ringing? Can you see the bell ringing? What does this tell you about both light and sound?

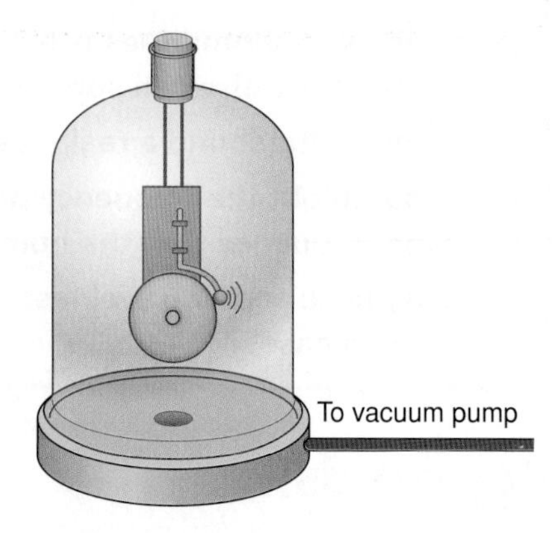

Figure 7.9 Bell jar attached to a vacuum pump.

Task

You are writing a section for the company brochure on the uses of light and sound.

You need to write reports on scientific investigations that you have carried out on the applications of light and sound waves.

Your task:

Step 1: Choose

Choose one application of light and one application of sound that you can carry out experiments on. There are some examples in the table below, but you can choose a different one if you prefer.

Light	Sound
plane mirror – the laws of reflection telescope – to explore space periscope – to see objects at a higher level cat's eyes – used in roads to reflect light concave lens – to correct vision convex lens – to correct vision another application of your choice	bell jar experiment – the need for a medium ultrasound – to scan the foetus in pregnancy ultrasound – to break down kidney stones sonar pulse – deep sea exploration and recovery another application of your choice

Step 2: Investigate

You need to investigate your chosen applications of light and sound. You must carry out a practical investigation for each one (so choose an application where you have access to the necessary practical equipment). Make sure each experiment has a clear aim and that you have all the equipment you need ready.

Step 3: Write up

Write up the results of your experiments. You must include the aim of the experiment and a labelled diagram of the equipment used. Record your results and draw a graph if appropriate. You may also have made a model of your application, for example a periscope.

Step 4: Present

Present your work as a report for the 'light and sound' company brochure. It should be interesting and appealing to the readers.

Make the grade

P3 For P3, you must complete this task:

Carry out an investigation into an application of the uses of waves.

Analysis

Essential science for M3

In this assignment you have to analyse the results of the investigation you carried out for P3.

When you analyse something you need to examine what you have investigated in detail to discover something or come to a conclusion. The **conclusion** could be a **pattern** or **trend** in the results demonstrated by a graph or some significant information.

Another engineer in your company has carried out an investigation on sound waves. You have been asked to look at her report and then draw a graph and provide a conclusion.

Which materials are best at absorbing sound?

Aim: To find out which materials are the best at **absorbing** sound.

Method: A sound is **transmitted** through each material and the speed of the sound is calculated and recorded in the table. Each material is tested three times.

Science: The faster the **speed of sound** in a material, the better it is at transmitting sound. The slower the speed, the better it is at absorbing sound. Engineers are interested in finding materials that are good sound-absorbers so that they can be used in buildings. We don't want to be able to hear through the walls in our homes and offices!

Results: The results of the investigation are shown in the table below.

Material	Speed of sound waves (m/s)			Average m/s	Sound-absorbing qualities
	Test 1	Test 2	Test 3		
Steel	5060	5089	5072	5074	Very poor
Glass	3962	3937	3945	3948	Fairly poor
Concrete	3403	3436	3424	3421	Fairly good
Plastic foam	456	435	443	445	Good
Rubber	156	167	159	161	Very good
Wood	3567	3544	3551	3554	Fairly good

Graph:

Conclusion:

Questions

1. Draw a graph showing the results of this investigation on sound waves. What type of graph should it be? A line graph? Or a bar chart? Use graph paper.
2. Which is the best sound-absorbing material?
3. What is the average speed of sound in this material?
4. Why did the engineer carry out each experiment three times?
5. Why do you think that materials like steel and glass are not very good at absorbing sound?
6. What do the good sound-absorbing materials have in common?
7. Why is it important for engineers to know which materials are good at absorbing sound?
8. Write a conclusion for this report. It should state clearly what the investigation has found out.

Key words:

pattern, trend, conclusion, absorbing, transmitting, speed of sound

Task

In addition to collecting data for your P3 investigations, you now have to analyse the results from both of your investigations.

Your task:

Step 1: Analyse the results

You will have tables of results for both of your investigations. These are your 'data' – lots of pieces of information. You need to find out if there is a pattern in the numbers of the data. Your brain is not like a computer; it does not deal with blocks of numbers very well. Your brain needs the blocks of numbers to be turned into pictures – such as line graphs, pie charts or bar charts. This is what you need to do to *analyse* your data. You need to choose what sort of graph or chart you might use to display your data.

Think of questions like: What are the variables I am comparing to get my data?

Was it a simple question like 'Have you felt dizzy after listening to very loud music?' You could analyse the responses to this with a pie chart. If the question was more complicated like 'What effect does thickness of padding have on sound transmitted to a sensor?' then you may need to use a line graph.

Step 2: Conclusion

Write a conclusion for each of your investigations.

What have you found out? Look at the patterns you have found in the numbers of your data. These may show a clear pattern that you can express as a sentence like 'Most males have felt dizzy after listening to loud music' or 'Thicker insulation cut down the sound reaching a sensor'.

Or could your conclusion be more detailed like '73% of males between 14 and 18 years old have been dizzy after loud music' or 'Doubling the thickness of insulation will halve the sound energy reaching an analogue sound sensor'.

Step 3: Present

You may wish to add these sections to the report you wrote for P3.

Make the grade

M3 For M3, you must complete this task:

Analyse the results of the investigation into the uses of waves.

More about light

Essential science for D3

Bending a pencil...

The **refraction** of light as it passes through different materials explains why the pencil in Figure 7.10 appears to bend in the water.

The speed of light is 300 million metres per second (m/s) in air, but in water it travels more slowly at about 225 million m/s. In glass its speed is different again – about 185 million m/s.

When light passes from air into water it changes speed. The sudden **change in speed** bends the light waves. You can see this effect by placing a pencil in a glass of water (see Figure 7.10) or shining a beam of light through a glass block (Figure 7.11).

Figure 7.10 Why does this pencil appear to bend?

Evaluation

When you evaluate your work you must look closely at what you have produced and be able to justify why you chose to do it that way.

In this section you will need to evaluate your investigations into the applications of light waves and sound waves, and try to suggest **improvements** to the **real-life applications**.

You can use these questions to help you plan your evaluation.

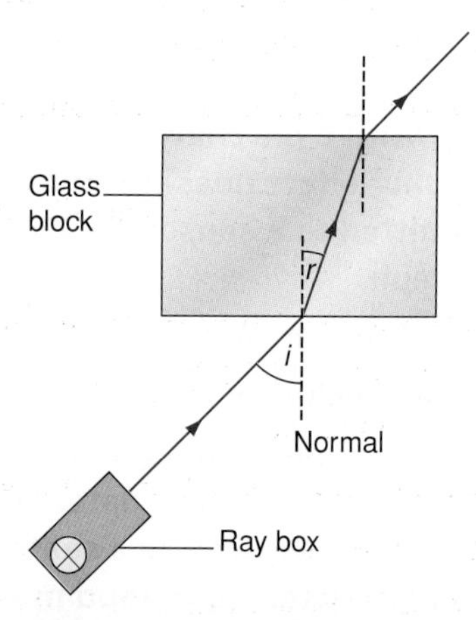

Figure 7.11 The refraction of light in a glass block. Notice how the light appears to bend as it enters the glass block. This is due to a change of speed.

Questions

1. What were you trying to achieve in this assignment?
2. Did you use scientific knowledge to plan your investigations?
3. Did you use scientific skills to carry out your investigations?
4. How would you know if your investigations were successful?
5. Were you able to make any firm conclusions?
6. If you were to change any part of your investigations, what would you change?
7. Are there any improvements you could make to your investigations?
8. What are the real-life applications of your chosen investigations?
9. Suggest improvements to the real-life applications that you chose to investigate.

Key words:

refraction, change in speed, improvements, real-life applications

Task

In addition to the investigation you carried out for P3 and the analysis you did for M3, you now have to evaluate both of your investigations and suggest some improvements to the real-life applications.

Your task:

Step 1: Evaluate your investigations

Look closely at the results (P3) and analysis (M3) for both of your investigations.

Use these questions to evaluate both of your investigations.

1. **What were you trying to achieve in this assignment?**
2. **Did you use scientific knowledge to plan your investigations?**
3. **Did you use scientific skills to carry out your investigations?**
4. **How would you know if your investigations were successful?**
5. **Were you able to make any firm conclusions?**
6. **If you were to change any part of your investigations, what would you change?**
7. **Are there any improvements you could make to your investigations?**
8. **What are the real-life applications of your chosen investigations?**
9. **Suggest improvements to the real-life applications that you chose to investigate.**

Step 2: Evaluate

Make an overall evaluation for your investigations.

Step 3: Present

Write up your evaluation. You may wish to add this section to the reports you wrote for P3 and M3.

When you have finished, show your teacher.

Make the grade

D3 For D3, you must complete this task:

Evaluate the investigation into the uses of waves in our world, suggesting improvements to the real-life application.

Unit 5

Chapter 8
Using electricity

Using electricity

Figure 8.1 Some security lights switch on automatically if movement is detected when it is dark.

Figure 8.2 Burglar alarms go off when an intruder activates a sensor.

Scenario

Electricity can do so much more for us than just boil the kettle when we want a hot drink. There are many different ways that electricity is used in modern society. In fact, the whole way our society runs is based on electricity - it is the energy source we use to make most things happen. Remember that nothing can happen without an energy transfer making it happen.

Electric circuits are our willing slaves, waiting patiently for us to turn them on. Sometimes we don't even need to do that - a lot of the jobs we get circuits to do are turned on automatically by sensors and control devices, without us having to think about it. Different components can be added to circuits to develop new devices that are activated by light or changes in temperature, eliminating the need for people to use switches to turn the devices on or off.

You are a technology technician. You need to design a device that can be turned on by light or changes in temperature. You will need to investigate how different components can be used and decide which system is the most appropriate one to use to do the job for you.

Grading criteria for Using electricity

To achieve a pass grade you need to:	To achieve a merit grade you also need to:	To achieve a distinction you also need to:
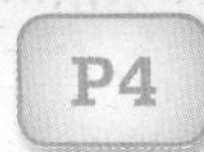 P4 carry out an investigation into an application of the uses of electricity	 M4 analyse the results of the investigation into the uses of electricity	 D4 evaluate the investigation into the uses of electricity in our world, suggesting improvements to the real-life application

Scenario

You are a nurse working in an intensive care unit. Hospitals use equipment that monitors the electrical activity in the body. There is a great deal of this sort of equipment being used in the health service. It relies on picking up tiny electrical signals from the body and amplifying them, or using sensors that make an electrical signal out of the body's functions.

Carry out an investigation into the uses of electrocardiograph (ECG) machines and defibrillators.

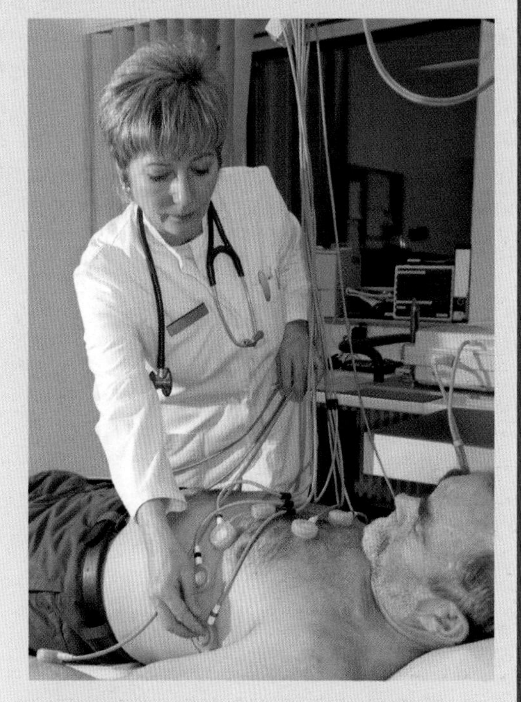

Figure 8.3 A nurse using an ECG machine to monitor a patient's heart.

Keeping it local

- Find out how the different sensors in automatic washing machines, dishwashers and other appliances in your home are used to keep the sequence of jobs in the correct order and happening for the correct amount of time.
- Speak to a local electrician about how they use electricity. Ask them to show you some different electrical components that they use. Do they look different to the ones that you use in school?

Careers

- Electrician
- Nurse
- Operating theatre technician
- Equipment designer
- Electronic technician
- Traffic signals engineer

Background science for the assignments

Light bulbs, buzzers and motors are all types of electrical components. They are transducers that can be inserted into electrical circuits to do various jobs. All circuit components transfer the electrical energy carried by the circuit to the surroundings as light, sound, movement, etc. Because they transfer energy, all the components have a **resistance** to the electric **current**.

Ammeters and voltmeters are measuring devices that are used to measure current and potential difference. Ammeters must be inserted in **series** with the other components. Voltmeters must be inserted in **parallel** across the components where they are measuring the potential difference.

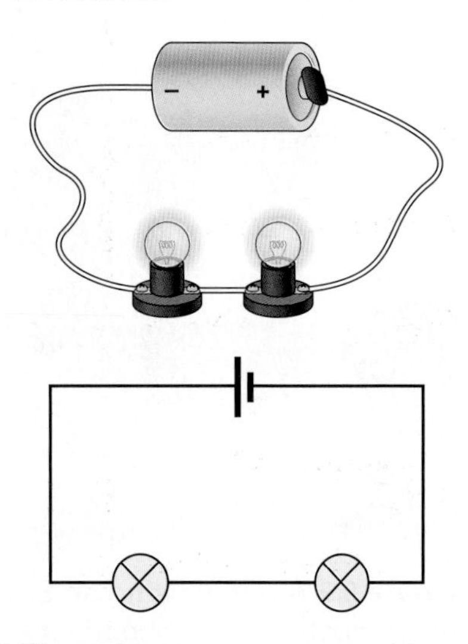

Figure 8.4 These lamps are arranged in series.

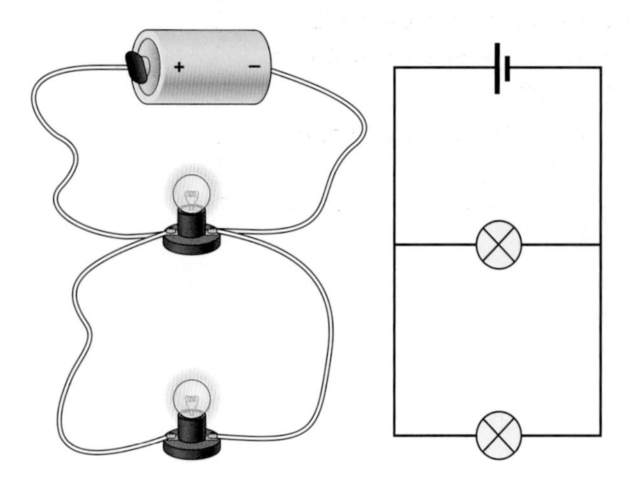

Figure 8.5 These lamps are arranged in parallel.

Ohm's Law

Ohm's Law was discovered in 1827 by Georg Ohm whilst he was doing experiments about electrical conductivity. It describes the relationship between current, **voltage** and resistance and is represented by the equation **V = IR**.

> **V = voltage:** The electrical potential energy, measured in **volts** (V).
> **I = current:** The number of electrons passing through the wire, measured in amperes or **amps** (A).
> **R = resistance:** The difficulty the electrons have in moving through the wire, measured in **ohms** (Ω).

Power and energy

Electrical devices also have a **power** rating. This shows how much electrical energy they transfer per second.

A bright light will transfer 100 **watts** of energy. This means that it transfers 100 **joules** of energy per second. Watts can also be calculated as current through the device multiplied by the potential difference across the component.

> 1 watt of power = 1 joule of energy per second
> watts (W) = amps (A) × volts (V)
> 1 kilowatt = 1000 watts

Energy use is calculated as joules of energy. A common unit of domestic energy used is the **kilowatt hour** – this is the energy transferred by a 1 kilowatt device in 1 hour.

> joules transferred (J) = watts × seconds the device is switched on

Key words:

amp, current, joule, kilowatt hour, ohms, Ohm's Law, parallel, power, resistance, series, volt, voltage, watt

Using electricity activities

Practical activities

1 Use the equipment provided to set up a series circuit similar to that shown in Figure 8.4. Insert an ammeter and a voltmeter. Find out:
   a) what the current is at different points in the circuit
   b) what the voltage is at different points in the circuit
   c) what happens to the current and voltage when you insert a second and then a third cell (battery).

2 Use the equipment provided to set up a parallel circuit similar to that shown in Figure 8.5. Insert an ammeter and a voltmeter. Find out:
   a) what the current is at different points in the circuit
   b) what the voltage is at different points in the circuit
   c) what happens to the current and voltage when you insert a second and then a third cell (battery).

3 What have you found out about current and voltage in series and parallel circuits?

Practise using Ohm's Law

4 Rearrange the equation $V = IR$ so that it can be used to calculate current when you know the voltage and resistance.

5 Rearrange the equation $V = IR$ so that it can be used to calculate resistance when you know the voltage and current.

6 Use the current and voltage readings that you collected in questions 1 and 2 to calculate the resistances in the circuits.

Electricity and the human body

7 a) Which of the materials in the list below are conductors of electricity?
   b) Which of the materials are insulators?

   copper wood aluminium cottonwool steel brass polythene rubber

   c) Explain your answers to parts a and b.
   d) Would you expect human tissue to be a conductor?

8 You may have seen (on TV) a defibrillator being used to restart a person's heartbeat. What do you think is happening? Why do the paramedics put damp pads on the person's body where they hold the 'paddles'?

9 Explain why a person can be electrocuted by the electricity from a pylon, but not electrocuted by the electricity from a torch battery.

Investigate!

Essential science for P4

Resistors are components in circuits that slow the flow of electric current. A bulb is an example of a resistor. Fixed resistors are used to reduce the current by a specific, fixed amount. Other special types of resistor are variable, enabling the size of the current to change in different circumstances.

Thermistors detect the temperature of the surroundings. This controls the resistance and therefore the size of the current.

Light-dependent resistors detect the amount of light in the surroundings. This controls the resistance and therefore the size of the current.

Diodes are components in circuits that ensure that the current only flows in one direction. There are various different types of diodes.

Light-emitting diodes are a special type of diode that emit light when a current flows through them.

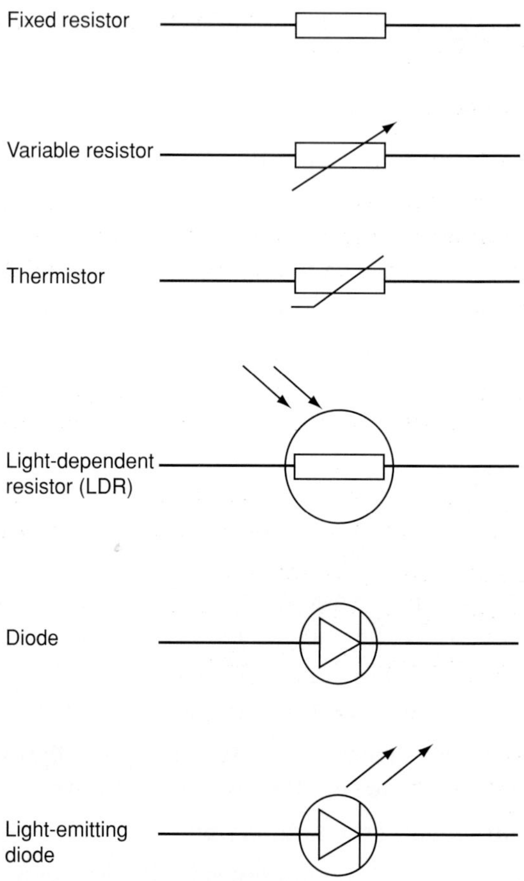

Fixed resistor

Variable resistor

Thermistor

Light-dependent resistor (LDR)

Diode

Light-emitting diode

Figure 8.6 Electrical symbols.

Electricity in the body

Electricity can be conducted through the human body. Machines can be used to monitor the electrical activity of the heart, and if the heart is not working correctly, electrical current can be delivered to it to shock it into resuming a regular heartbeat.

An **electrocardiogram** (ECG) machine measures the electrical activity in the heart. A **defibrillator** machine is used to restart a person's heart if it is not beating regularly.

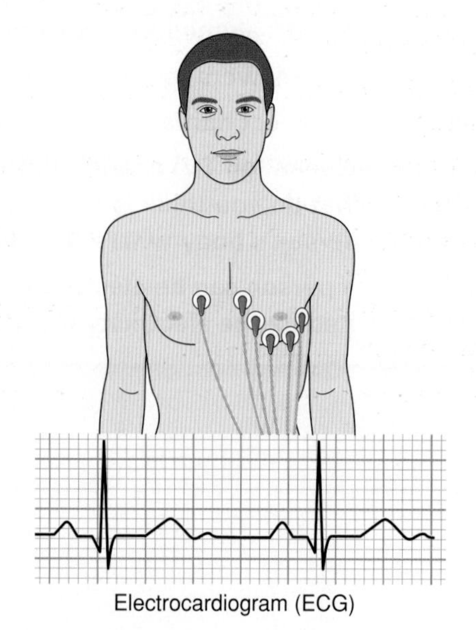

Figure 8.7 A patient having an ECG.

Key words:

defibrillator, diodes, electrocardiogram, light-dependent resistors, light-emitting diodes, resistors, thermistors

Task

You are going to investigate how different electrical components affect the current in a circuit. This will help you to design your own device that can be turned on by changes in light or temperature.

Your task:

Step 1: Choose

Choose an application of the uses of electricity (for example burglar alarm, night light, smoke alarm, etc). What electrical components does your chosen device consist of? How do the components affect the electrical current that flows?

Step 2: Measure

Set up simple circuits that include the same components as the application that you have chosen. Measure the current and voltage in different cosnditions. As you do your practical work, think about the method that you are using. Will your results be accurate? Will they be reliable and free from errors?

Step 3: Record your results

Record your results in a chart or table.

Here is a sample results table:

	Light level	Voltage	Current
Attempt 1			
Attempt 2			
Attempt 3			

Tip
It is likely that you will have to draw multiple results tables!

Look at the sample results table and ask yourself the following questions:

- **Would this results table encourage you to collect reliable results?**
- **What range of light levels or temperatures would be appropriate to use?**
- **How many times should you try each different light level or temperature?**
- **Does it matter how quickly you repeat the experiment at each light level or temperature?**

Make the grade

P4 For P4, you must complete this task:

Carry out an investigation into an application of the uses of electricity.

Analyse!

Essential science for M4

When you analyse the results of an investigation you need to look at the method you used and the results you obtained and draw conclusions.

Graphs are a good way to display results as they help you to see trends and patterns.

- You can plot the size of the current flowing through a circuit at different light levels or temperatures.
- You can calculate the resistance in the circuits if you have recorded both the current and the voltage in the circuits.

> Remember **Ohm's Law**:
>
> $$\text{resistance} = \frac{\text{voltage}}{\text{current}}$$

The shape of the graphs you draw for your investigation will vary depending on what type of resistor is used. If you have a **line of best fit** that is straight, you know that the current and voltage are directly **proportional** to each other and that the resistance remains constant. If the line of best fit is a curve, the resistance changes as the voltage is changed.

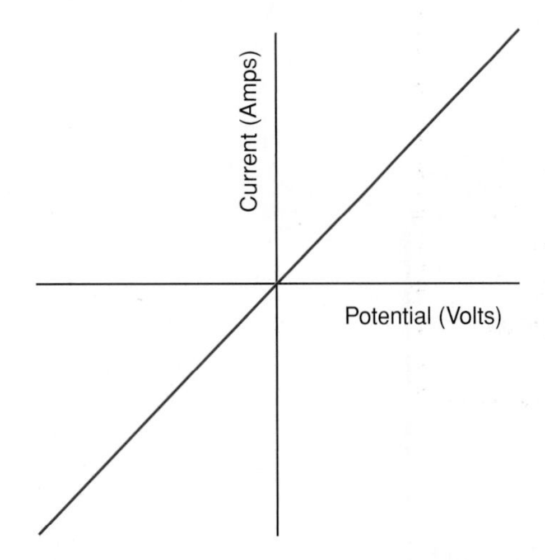

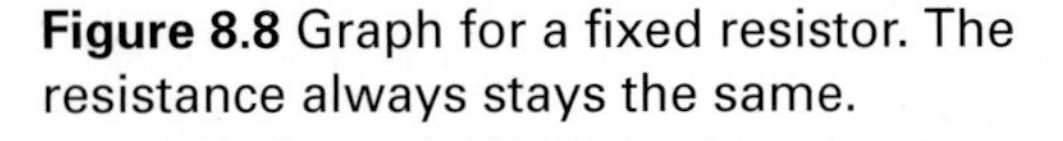

Figure 8.8 Graph for a fixed resistor. The resistance always stays the same.

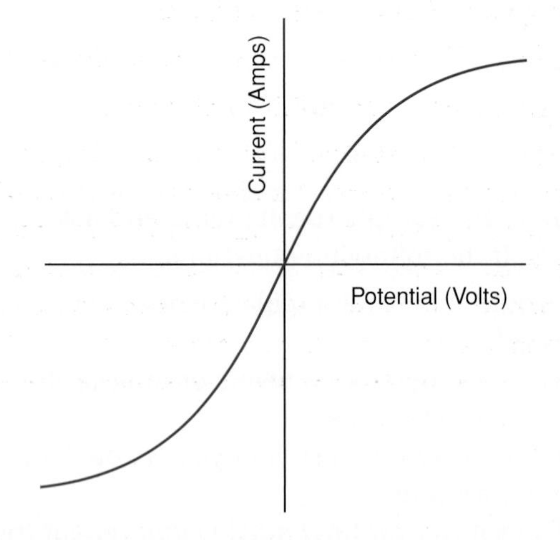

Figure 8.9 Graph for a filament bulb. The resistance is higher in the steeper sections of the graph.

Electrical resistance refers to the electrons' ability to flow through the circuit. The greater the resistance, the harder it is for a current to flow at a given voltage. The atoms in the material that makes up the circuit affect the resistance; if the nuclei of the atoms are behaving in a way that obstructs the flow of electrons, the resistance is increased.

Key words:

Ohm's law, line of best fit, proportional

Task

You now need to analyse the results of the investigation you carried out for P4.

Your task:

Step 1: Plot graphs

Plot graphs to show how the current and voltage changed in your circuit when you used different conditions. When you plot your graphs, make sure that they all have:

- **a clear title related to the investigation you carried out**
- **scales on the axes that will spread your data most of the way across or up the graph grid**
- **units for the measurements you used.**

Line graphs should have one clear line that goes close to all the points. This should be a straight line or a smooth curve.

What patterns do you notice in your results?

If you used a light-dependent resistor:

- **How did the amount of light affect the current that flowed?**
- **What does this tell you about the resistance in the circuit?**

If you used a thermistor:

- **How did the temperature affect the current that flowed?**
- **What does this tell you about the resistance in the circuit?**

If you used a light-emitting diode:

- **How did it affect the resistance in the circuit?**

Step 2: Provide a scientific explanation

Using your knowledge of electric current, explain why the resistance in your circuits changed. You should include a brief description of how the components that you have used in your circuit work. You do not need to explain how the actual components work in detail. Just explain how their change in resistance affects the behaviour of the circuit.

Make the grade

For M4, you must complete this task:

Analyse the results of the investigation into the uses of electricity.

Evaluate!

Essential science for D4

When you have completed your experimental work and decided on your conclusion, you need to look back over what you did in a self-critical manner. This is part of the way you get the best possible answer to your original question for the investigation. There are two sides to evaluating a scientific investigation:

- Were your methods the best you could manage, or could you make your experiments better?
- Did you produce data that gave a clear consistent pattern that you could explain, or were there some doubts about the quality of the data and conclusion?

You need to make decisions about the suitability of the method or equipment used and suggest how improvements could be made. Sometimes the accuracy of the measuring instruments could be improved, or perhaps your own technique for taking readings at regular intervals made the results inaccurate and led to unreliable data.

Errors and anomalous results

There are many ways that errors can occur. A **systematic error** is one that happens every time a procedure is carried out. These errors can be difficult to spot; you may seem to have a set of accurate results, but they are not close to the actual value. A **random error** is easier to spot as it produces an **anomalous result**. If you have anomalous results then you should either repeat that part of the experiment or disregard the anomalous result.

A random error will often stand out on your graph as an anomalous result, but a systematic error is much more difficult to spot as it may shift all the graph points up or down by the same amount.

Look critically at how well your graphs give lines or patterns that are consistent and not variable. Do your graph points fit closely to a straight line or curve, or are they either side of it?

Investigations must produce **valid** results. They must be **reliable** and **fit for purpose**. This means that the data collected from small-scale laboratory investigations can be used to design and improve real-life products and applications.

Key words:

anomalous result, fit for purpose, random error, reliable, systematic error, valid

Task

You now need to evaluate your investigation.

Your task:

Remember that evaluation is about:

- the quality and carefulness of your practical methods
- how well your data fit a pattern that you can explain by using scientific theories about the way electrical circuits work.

Remember to use technical words correctly:

- The potential difference between two points is measured in volts.
- The current flowing in part of a circuit is measured in amps.
- Resistance is the ratio of volts (potential difference) to amps (current) in a component in a circuit (volts divided by amps).

Step 1: How good are your results?

- Look at your results. This is best done by looking at your graphs. Your mind can analyse pictures better than numbers.
- Do you have any anomalous results? Do any results stick out from the pattern on your graphs?
- What errors may have caused them?
- Think about your method and the equipment that you used. What systematic errors may have occurred?
- Were the measuring instruments you used accurate enough?
- Are your results repeatable? Do they all fit the pattern of the graph?
- Are your results fit for purpose? Are you able to explain the pattern and conclusion you found by using a scientific idea?

Step 2: Suggesting improvements

- What is your real-life application?
- How do your results tell you about the problems you might find in designs for the real-life application?
- What improvements could be made to the real-life application as an outcome of the results that you have collected?

Make the grade

For D4, you must complete this task:

Evaluate the investigation into the uses of electricity in our world, suggesting improvements to the real-life application.

Unit 6

Chapter 9
Healthy living

A better lifestyle

Figure 9.1 Health spas are becoming more and more popular.

Health spas have become very popular. They provide a place where people can relax, meet up, exercise and eat healthy meals. Some can even be thought of like medical facilities giving healthcare advice.

Scenario

You are a health and fitness coach working at a famous health spa. Your job is to help people understand how diet and exercise affect the human body. You have a lot of important people under your care who depend on you to keep them healthy and active so they can do high-powered jobs.

Part of your role is to design diet plans (menus) and exercise plans to promote healthy living. You include a lot of lifestyle advice.

Your diet plans and exercise plans have to be tailored to suit the needs of different people who live different lives. You have to anticipate their needs and the demands their lives place on them. In particular, you have to keep them ready for anything and looking good.

Grading criteria for Healthy living

To achieve a pass grade you need to:	To achieve a merit grade you also need to:	To achieve a distinction you also need to:
P1 assess the possible effects of diet on the functioning of the human body	M1 explain how the diet and exercise plan will affect the functioning of the human body	D1 evaluate your exercise plans and justify the menus and activities chosen
P2 design diet and exercise plans to promote healthy living		

Keeping it local

- You must have a fitness gym or health spa near you – probably several of them. Try to visit one and see what kind of facilities they have to help people keep fit and healthy.
- Carry out a survey on the subject of exercise on the people in your class. Find out:
  - how many people exercise regularly
  - how many think they do not do enough exercise
  - how many think they eat a healthy diet
  - how many made the New Year's resolution to eat more healthily and exercise more.

Figure 9.2 Thousands of people make it their New Year's resolution to 'Get fit, lose weight and become healthier'.

Careers

- Fitness trainer
- Sports therapist
- Dietician
- Nutritionist
- Physiotherapist
- Food technician
- Slimming club advisor

Scenario

You are a medical technician in the armed services. You are working on the rehabilitation of injured men and women recovering from their injuries. The injured people need a specific programme to rehabilitate them. Often this requires a programme of exercise for their injured and damaged limbs.
Also a specific diet is needed to build up their bodies specifically to speed recovery.

Background science for the assignments

Being healthy is about more than just eating healthy food. It includes exercising, relaxing, and not being too stressed. In this chapter you will look at how people can develop diseases through poor diet and lack of exercise.

Types of disease

Mental: Things that affect the mind, such as depression, anorexia nervosa, phobias, bipolar disorder or schizophrenia.

Non-infectious: Diseases such as diabetes, cancer or heart disease are non-infectious – they cannot be caught from another person.

Physical: Diseases that affect the physical functioning of the body or its parts, e.g. heart disease, arthritis, obesity, reduced mobility, sports injuries, stress or cardiovascular and respiratory diseases.

Infectious: Diseases can be infectious (can be caught from another infected person), e.g. influenza, athlete's foot, malaria, polio, measles, mumps, rubella, typhoid, food poisoning or SARS.

A well-balanced diet must contain seven essential **nutrients** (nutrients are the useful substances in our food). These are:

1. carbohydrates (bread, cereals, sugar, etc.)
2. proteins (meat, fish, eggs, beans, etc.)
3. fats (oils, butter, margarine, etc.)
4. fibre (brown bread, brown rice, etc.)
5. vitamins (A, B1, B2, C, D, etc.)
6. minerals (calcium in cheese, iron in meat, etc.)
7. water.

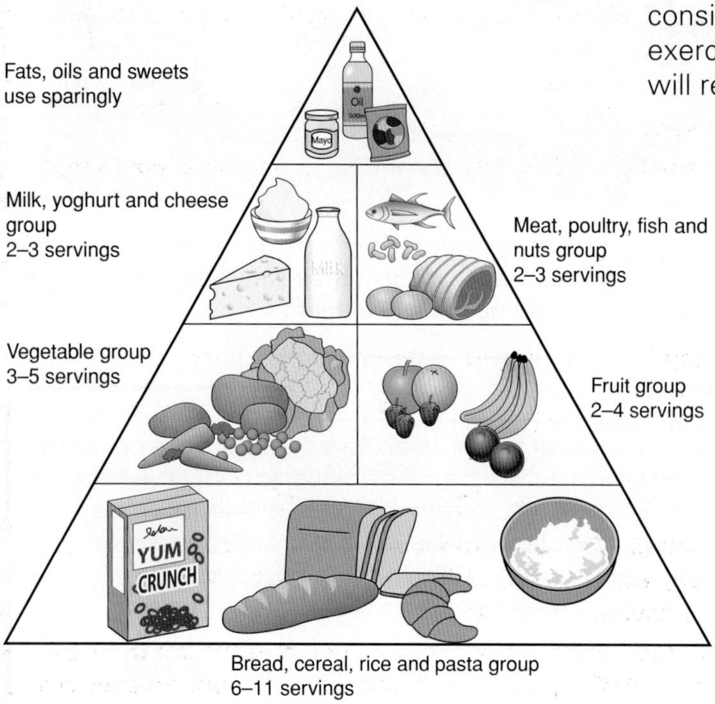

Fats, oils and sweets use sparingly

Milk, yoghurt and cheese group 2–3 servings

Meat, poultry, fish and nuts group 2–3 servings

Vegetable group 3–5 servings

Fruit group 2–4 servings

Bread, cereal, rice and pasta group 6–11 servings

Figure 9.3 A food pyramid.

The food pyramid in Figure 9.3 shows the proportions that you should eat these foods in to have a well-balanced diet. You need plenty of carbohydrates, some proteins and a small amount of fat. You should also avoid having too much salt in your diet.

However, the ideal proportions of each nutrient group will vary from person to person, for example a teenager who is still growing will need a higher proportion of protein in their diet than an adult will. Every individual also needs to consider the amount of exercise they do. If you exercise a lot, you use more energy and so you will require more food.

Food allergies

Unfortunately, some people develop **allergies** to certain foods. Peanut and nut allergies are quite common and some people (coeliacs) are allergic to gluten, which is found in foods like wheat, rye and barley.

Question

1 Can you think of any other food allergies that people have?

Key words:

allergy, infectious, mental, non-infectious, nutrient, physical

Healthy living activities

1 What are the seven nutrients that are needed to maintain good health?

2 Copy and complete the table below by choosing foods from the following list that are high in each nutrient. Each food can be used *more than once*, for example brown rice is high in fibre *and* carbohydrates.

Food list: red meat, eggs, tuna fish, pasta, fried chicken, grilled chicken, cabbage, lentils, porridge, chips, chocolate, oranges, crisps, pizza, cereals, cheese, sweetcorn

Nutrient	Types of food
carbohydrate	brown rice
protein	
fat	
vitamins	
minerals	
fibre	brown rice
water	

3 Vitamins are essential for the body as they help to speed up chemical reactions. Try to find out which diseases you can develop if you lack:

a) vitamin C

b) vitamin D.

4 Minerals are needed by the body to make tissues and also in chemical reactions.

a) Which mineral is important for healthy teeth and bones?

b) Which mineral is needed to make red blood cells?

5 Fats are also important, although too much fat is not good for us. Fat provides a good store of energy. When might this store of energy be useful?

6 Why is it important to have a diet that is high in fibre?

7 Give three foods that contain a large amount of fibre.

8 Heart disease can be caused by having a poor diet. Which types of food in particular could lead to a person developing heart disease?

9 What are the effects of:

a) over-eating

b) under-eating?

10 Copy the table below and put the following lifestyle choices in the correct column:

Good for the body	Bad for the body

Word list: eating lots of vegetables, smoking, adding salt to food, meditation, plenty of exercise, drinking alcohol in moderation, lack of sleep, eating lots of sweets, taking recreational drugs, drinking lots of coffee, having a stressful job

11 **Practical activity:** There are well-known food tests that you can carry out in the lab to test for certain nutrients. Carry out the Food Test Activity on a range of foods that your teacher will provide for you. Draw up a table of results.

- Fats: emulsion test
- Proteins: biuret test
- Starch (carbohydrates): iodine test
- Sugar (carbohydrates): Benedict's test

The human body

Essential science for P1

How diet can affect the human body

A healthy diet contains seven nutrients: carbohydrates, fats, proteins, vitamins, minerals, fibre and water. If a person does not get the right amounts of these nutrients in their diet it can affect the body in various ways. You can end up being **malnourished** if your diet does not supply you with the right amounts of nutrients, and this can lead to diseases developing.

Food contains energy and different types of foods have different energy values. These are usually detailed in the 'nutrition information' on food packaging (see Figure 9.4).

If you get too much energy from your food, you may put on weight or even become obese. If you don't get enough energy from your food, you could become too thin to be healthy and will lack vital nutrients. The energy in food is often measured in calories but the correct unit is kilojoules (kJ).

There are many diseases and conditions that can result from a poor diet, for example:

- too much fat: obesity, arthritis, **heart disease**
- not enough protein: muscle weakness, poor growth
- not enough carbohydrate: lack of energy
- too much carbohydrate: extra energy stored as fat
- not enough vitamins: diseases like scurvy (lack of vitamin c) and rickets (lack of vitamin d)
- not enough minerals: anaemia (lack of iron), brittle bones (lack of calcium)
- not enough fibre: constipation, infections of the intestines
- not enough water: dehydration.

A healthy diet means not eating too much salt, sugar or saturated fats. Alcohol should be consumed only in moderation. Foods treated with pesticides and other chemicals should be avoided, and so should highly processed foods containing a lot of additives and preservatives.

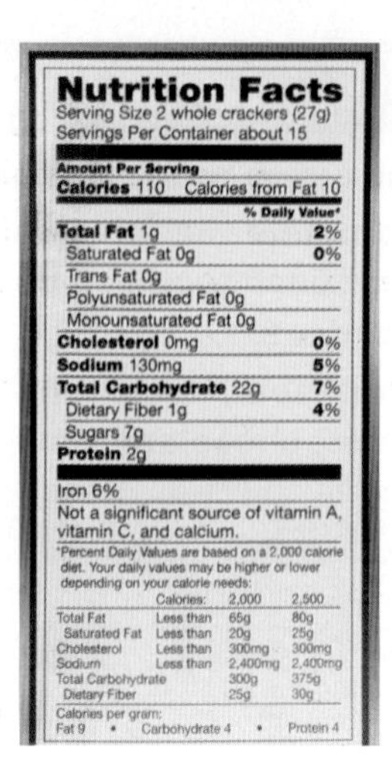

Nutrition Facts
Serving Size 2 whole crackers (27g)
Servings Per Container about 15

Amount Per Serving
Calories 110 Calories from Fat 10

	% Daily Value*
Total Fat 1g	**2%**
Saturated Fat 0g	**0%**
Trans Fat 0g	
Polyunsaturated Fat 0g	
Monounsaturated Fat 0g	
Cholesterol 0mg	**0%**
Sodium 130mg	**5%**
Total Carbohydrate 22g	**7%**
Dietary Fiber 1g	**4%**
Sugars 7g	
Protein 2g	

Iron 6%

Not a significant source of vitamin A, vitamin C, and calcium.

*Percent Daily Values are based on a 2,000 calorie diet. Your daily values may be higher or lower depending on your calorie needs:

	Calories:	2,000	2,500
Total Fat	Less than	65g	80g
Saturated Fat	Less than	20g	25g
Cholesterol	Less than	300mg	300mg
Sodium	Less than	2,400mg	2,400mg
Total Carbohydrate		300g	375g
Dietary Fiber		25g	30g

Calories per gram:
Fat 9 • Carbohydrate 4 • Protein 4

Figure 9.4 Nutrition information on food packaging.

Effect on the heart

A healthy heart is important for everyone! The heart provides the body with blood, which contains oxygen and the dissolved food substances we need to live. If the heart becomes affected by a poor diet, it will affect the way the body works.

The heart can be badly affected by a high-fat diet because the **coronary arteries** can get blocked up with fatty deposits from the food. This may result in blockages that can lead to heart disease, and if the blockage is severe it can cause a heart attack!

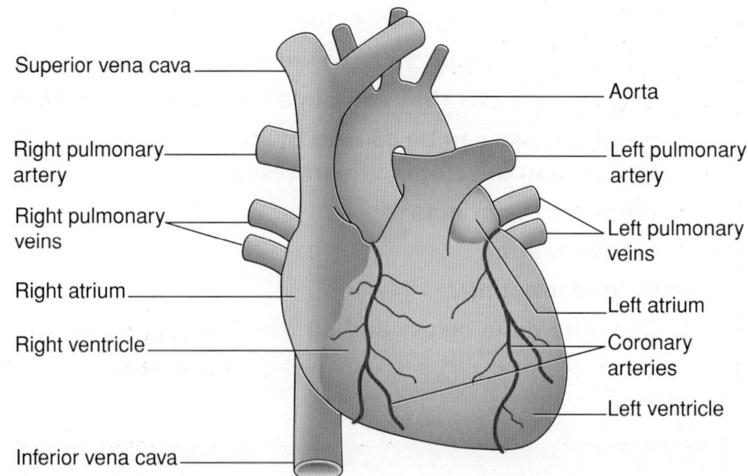

Figure 9.5 The human heart. The heart can be affected by a poor diet, especially if it is a high-fat diet.

Key words:

coronary arteries, heart disease, malnourished

Task

You are going to launch a new diet plan for a group of people who want to lose weight and get fitter. You will need to tell them how the different food groups affect the body and how over-eating can lead to them putting on weight. You will also need to warn them about the diseases that can develop as a result of having a poor diet.

Your task:

Step 1: Functions of the food groups

You need to state what the main function of each food group is in a healthy diet. Remember to include:

- **carbohydrates**
- **proteins**
- **fats**
- **vitamins**
- **minerals**
- **fibre**
- **water.**

Step 2: Choose diets

Choose four types of diet from the list below and explain how this diet can affect the functioning of the human body. You must say what will happen to the human body and mention any diseases that may result from eating this type of diet over a long period of time.

Diets that contain:

- **too much fat**
- **too little protein**
- **too little carbohydrate**
- **too much carbohydrate**
- **too few vitamins**
- **too few minerals**
- **too little fibre**
- **too little water.**

Step 3: Present

You must present your work to a group of people, so it would be best to present it as a PowerPoint presentation or a set of posters.

Make the grade

P1 For P1, you must complete this task:

Assess the possible effects of diet on the functioning of the human body.

Diet and exercise plans

Essential science for P2

Energy requirements

Different people require different amounts of energy. For example:

- A typical teenage boy with a mass of 50 kg requires about 6550 kJ of energy per day.
- An average adult woman who has a mass of 65 kg requires about 8520 kJ of energy per day.
- An elderly person living a retired lifestyle needs 5040kJ of energy per day

As a general rule, for every kilogram of body weight, 5.46 **kilojoules** are required every hour.

So for the teenage boy: 50 kg × 5.46 kJ = 273 kJ per hour.

273 × 24 hours = 6550 kJ needed per day.

However, the amount of energy required will also depend on how active the person is. If they are very active they will use up more energy so they will need to eat more food to get the energy they need.

The energy we take in has to be blended. A good blend of the different food types would be:

- 57% **carbohydrates**
- 30% **fats**
- 13% **protein**

So the teenage boy needs:

- 57% of his 6550 kilojoules from carbohydrates = **3734 kilojoules**
- 30% of his 6550 kilojoules from fats = **1965 kilojoules**
- 13% of his 6550 kilojoules from proteins = **851 kilojoules**

Exercise plans

It is advisable for all people to exercise every day – but this does not need to be strenuous exercise at the gym. Even a 20-minute walk every day can have beneficial effects on your health.

Key words:

carbohydrate, fat, kilojoule, protein

Weekly exercise plan
Monday: *30 minute walk*
Tuesday:
Wednesday: *dance club*
Thursday: *30 minute walk*
Friday: *30 minute walk*
Saturday: *yoga class*
Sunday:

Heart activity

Figure 9.6 is a diagram of the heart. Match the numbers (1–9) to the correct labels (A–I):

A blood to body
B valves
C right ventricle
D left ventricle
E blood from body
F right atrium
G left atrium
H blood from lungs
I blood to lungs

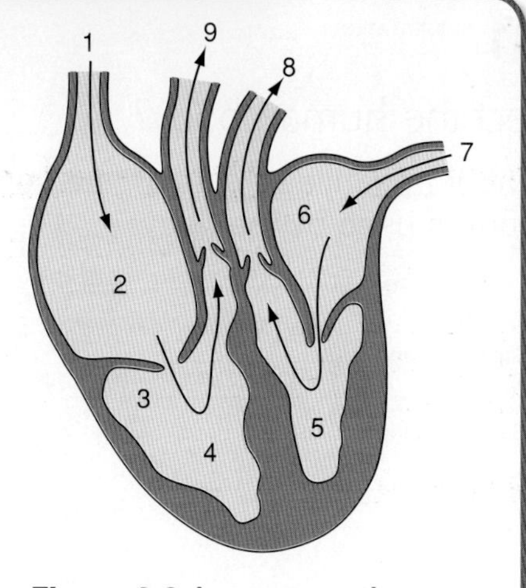

Figure 9.6 A cross-section through the heart.

Task

As a health and fitness instructor you need to be able to design diet plans (menus) and exercise plans for different types of people.

Your task:

Step 1: Choose

Choose three clients who you will design diet and exercise plans for.

The first must be a teenager like yourself (they could be an active sport/dance person or a couch potato) you decide.

The second should be an elderly person (they could be active or living a sedentary life).

For the third choose from this list:

- A pregnant woman
- An overweight gentleman aged 45 who is NOT very active
- A professional football player
- A call centre worker
- A named person you know – your friend, mum, dad etc.

Step 2: Design diet plans (menus)

For each person you chose in Step 1, design a diet plan covering 1 week. The diet plan must contain a balance of all the main nutrients and must be specifically designed for the person in mind. You can use the menu diary planning sheets to help you.

Each day should include breakfast, lunch, dinner and snacks. You need to say why you have chosen each of the foods you include in the plan. Make the food interesting and varied over the week.

Step 3: Design exercise plans

You must now design an exercise plan for each of your two chosen people. The chosen exercise for each day can be added to the menu sheets or the exercise plan can be done separately.

When you have finished, show your teacher.

Make the grade

P2 For P2, you must complete this task:

Design diet and exercise plans to promote healthy living.

Diet, exercise and the human body

Essential science for M1

How do diet and exercise affect the human body?

Food and exercise can have positive or negative effects on the body. This is how each of the main nutrient groups is used by the body:

Key words:

digestive system, fibre, heart, minerals, pulse rate, vitamins

- **Proteins:** Essential for growth and repair.
- **Fats:** A source of energy and important for keeping warm and insulation.
- **Carbohydrates:** Our main source of energy.
- **Minerals**: Needed for body tissues and for chemical reactions.
- **Vitamins**: Play important roles in many chemical processes in the body.
- **Water:** Essential for normal body functioning. It helps to carry other nutrients around the body.
- **Fibre**: The fibrous indigestible portion of our diet is essential for a healthy **digestive system**.

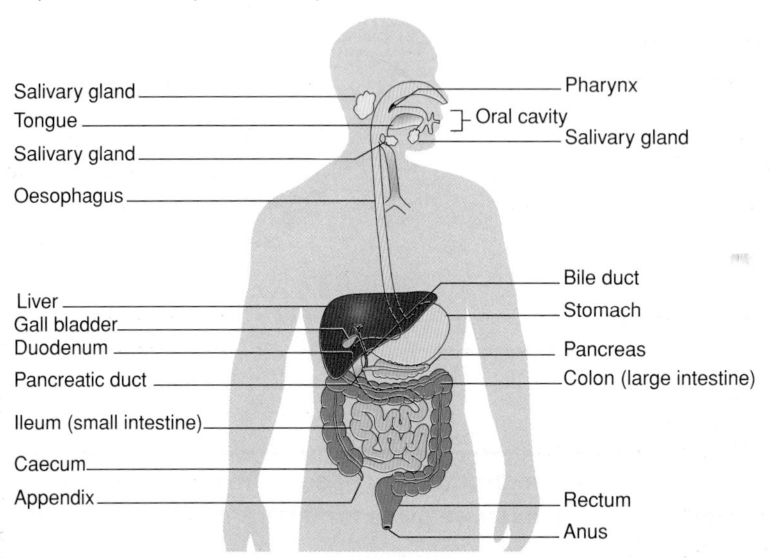

Figure 9.7 The organs of the digestive system.

Questions

1 Carry out research to find out how the diet affects the following parts of the body:
   a) the heart
   b) the blood vessels
   c) the digestive system (in particular the small and large intestines)
   d) the muscles and body tissues.

2 Carry out research to find out how exercise affects the following parts of the body:
   a) the heart
   b) the blood vessels
   c) the lungs
   d) the muscles.

Practical activity

What effect does exercise have on the heart (circulatory system)?

- **Take your pulse rate at rest for 1 minute and record the result in beats per minute.**
- **Carry out low-level exercise for 2 minutes (walking on the spot). Measure and record your pulse rate after the exercise.**
- **Carry out medium-level exercise for 2 minutes (jogging on the spot). Measure and record your pulse rate after the exercise.**
- **Carry out strenuous exercise for 2 minutes (star jumps, sprinting). Measure and record your pulse rate after the exercise.**

Copy the table below and record all your results in it.

	At rest	Low-level exercise	Medium-level exercise	Strenuous exercise
Pulse rate at rest (beats/min)				
Increase in heart rate from rest (beats/min):	–			

Conclusion:

What did you find out? How does exercise affect the human circulatory system?

Nutrients activity

Match each nutrient in List A to its correct use in List B.

List A	List B
fats	needed for body tissues and chemical reactions
proteins	helps to carry other nutrients around the body
minerals	the main source of energy
carbohydrates	for growth and repair
vitamins	helps to keep food moving through the gut
fibre	keep the body warm by insulating it
water	have an important role in many chemical reactions

Task

Now that you have designed diet plans (menus) and exercise plans for different people, you need to explain to them how the plans will affect their bodies and what benefits they will see if they follow your plans.

Your task:

Step 1: Describe

Describe in more detail the three people that you designed exercise and diet plans for in P2.

- **What were they hoping to achieve by following your plans?**
- **Were they hoping to lose/gain weight, become fitter or did they just want to stay healthy?**
- **How long should they follow your diet and exercise plans for?**

Step 2: Explain

For each of your three clients, explain how their diet and exercise routines will affect their body.

For example, one effect of the diet might be:

If they have lots of carbohydrates, this will give them plenty of energy for life processes such as respiration.

Think about which parts of the body will be most affected by the exercise and diet plans.

Step 3: Present

You will need to write a report to explain how your diet and exercise plans will affect the functioning of the human body. It will be useful to include some diagrams in your report.

When you have finished, show your teacher.

Make the grade

M1 For M1, you must complete this task:

Explain how the diet and exercise plan will affect the functioning of the human body.

Evaluation

Essential science for D1

When you **evaluate** your work you must look closely at what you have produced and be able to justify why you chose to do it that way. In this section you will need to evaluate all of the **exercise plans** and **diet plans** (menus) that you have produced. One way to do this is to look at what you were trying to achieve and see if you achieved it successfully or not.

You can use the following questions to help you.

1. What were you trying to achieve in this assignment?
2. What do you think your 'client' wanted to achieve?
3. Did you use scientific knowledge to plan your menus?
4. Did you use scientific knowledge to plan your exercise plans?
5. How would you know if your diet plans were successful?
6. How would you know if your exercise plans were successful?
7. Are there any **improvements** you could make to your diet and exercise plans?

Figure 9.8 Sarah before and after following healthy diet and exercise plans.

This is Sarah. She wanted to lose weight and become healthy. She followed the diet and exercise plans prepared for her by the health and fitness coach at her health spa. She had to follow the diet and exercise plans for a long period of time. It took one year for her to achieve her target weight.

Questions

1. Do you think that Sarah's efforts have been successful?
2. What questions would you like to ask her? Think of at least three.
3. What other benefits may there have been to Sarah's health, aside from the weight loss?
4. Do you think that Sarah looks happier?
5. Now that she has reached her ideal weight, what diet and exercise advice would you give to Sarah to help her maintain her new weight?

Key words:

diet plan, evaluate, exercise plan, improvement

Task

As a dietician and fitness coordinator you have already designed diet plans (menus) and exercise plans for different types of people in P2. You have also explained how the diet and exercise plans affect the body in M1. You are now going to evaluate the work you produced in P2 and M1.

Your task:

Step 1: Evaluate

For each of your clients, you need to look closely at the plans you produced for them and answer these questions honestly:

1. **What were you trying to achieve in this assignment?**
2. **What do you think your 'client' wanted to achieve?**
3. **Did you use scientific knowledge to plan your menus? Give some evidence of this.**
4. **Did you use scientific knowledge to plan your exercise plans? Give some evidence of this.**
5. **How would you know if your diet plans were successful?**
6. **How would you know if your exercise plans were successful?**
7. **Are there any overall improvements you could make?**
8. **Explain particularly why the diet solutions for the teenager and the elderly person are quite different.**

Step 2: Present

Present your evaluation as a report.

When you have finished, show your teacher.

Make the grade

D1 For D1, you must complete this task:

Evaluate your exercise plans and justify the menus and activities chosen.

Unit 6

Chapter 10
Staying healthy

Combating cancer

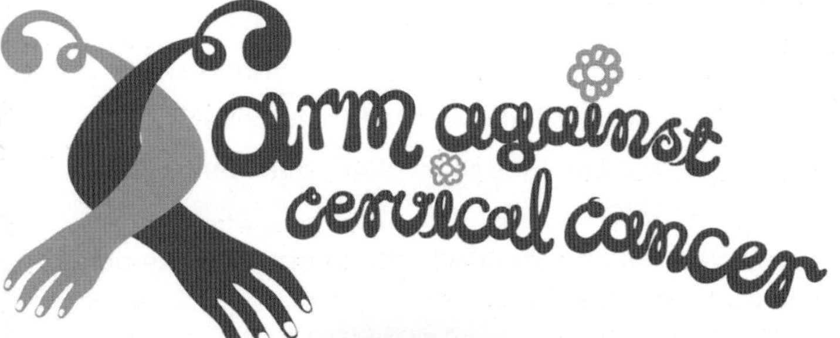

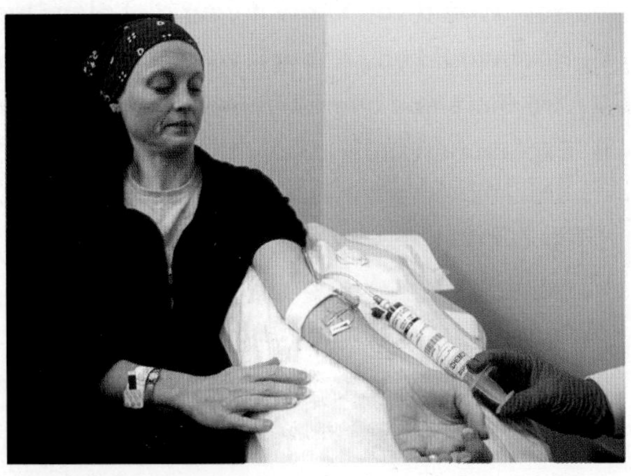

Figure 10.1 Despite the best treatment available, Jade Goody died from cervical cancer in March 2009.

Scenario

There is a lot of work being done in the field of cancer research. Cancer has a multitude of forms and is a killer disease, affecting some young people as well as older adults. The causes of cancer are still mostly unknown and poorly understood, but lifestyle choices (like smoking) can play a big part. People can be educated to help them reduce their risk of getting cancer by changing their lifestyle. However, there is no way to reduce the risk to zero.

A huge amount of research has been done into the prevention of cervical cancer. You have been employed by the charity Cancer Research to raise awareness of how the body works to defend itself against the micro-organisms that can cause cancer. You will be speaking to teenage girls to make them aware of what they should do to help protect themselves against cervical cancer.

Grading criteria for Staying healthy

To achieve a pass grade you need to:	To achieve a merit grade you also need to:	To achieve a distinction you also need to:
P3 outline how the immune system defends the human body	**M2** describe the action of each component of the immune system	**D2** evaluate the effectiveness of vaccination and screening programmes
P4 identify the role of specific health screening programmes	**M3** describe the changes in the human body following a vaccination	

Scenario

Every year, people are struck down by flu. Flu is a disease transmitted by micro-organisms. The micro-organisms invade the body and the immune system struggles to cope.

Most people can fight off this attack on the body. In fact, most people will recover from flu if they look after themselves properly while they are ill. However, some people cannot recover easily and become very ill. There are treatments such as flu vaccinations and antiviral drugs that can help these people cope with flu, but they are expensive.

You have been employed by the NHS to spread awareness about how the flu virus affects the body and what we can do to fight it off. You need to inform people about how to beat flu, and you need to inform the 'at risk' groups about treatments available to help with combating flu.

Keeping it local

- Have a chat with your local GP about the different vaccinations available.
- Speak to a vet about the vaccinations that your pets need. Do they work in the same way as vaccines for humans do?
- Carry out a survey to find out people's opinions about the MMR vaccination. Did you and your friends have it? Try to find out why some parents decided not to let their children be vaccinated.

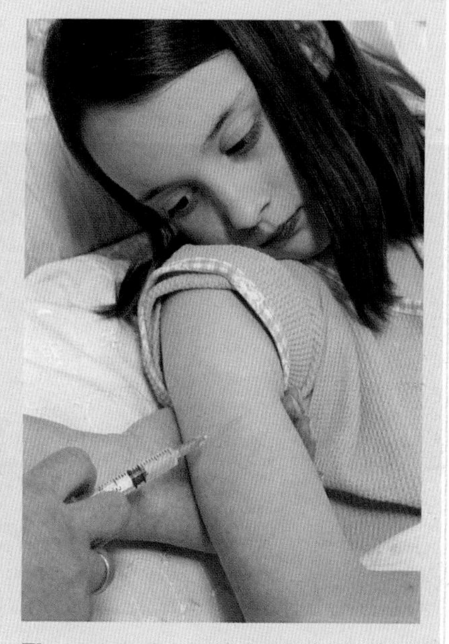

Figure 10.2

Careers

- Nurse
- Doctor
- Health advisor
- Hospital laboratory technician
- Hospice worker
- Charity worker

Background science for the assignments

Organisms are living things. Some are microscopic, while others are larger and have evolved to have **complex** systems that are essential for life. The human body is complex and has many specific requirements.

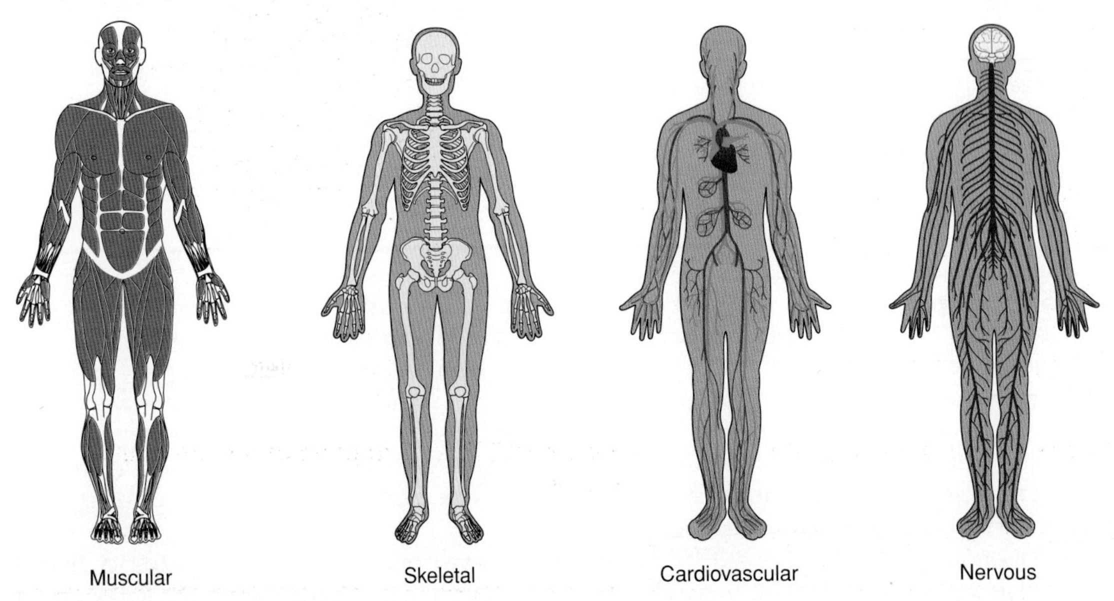

Figure 10.3 The human body consists of different systems working together. These diagrams show the muscular, skeletal, cardiovascular and nervous systems.

Unfortunately, the body systems do not always work correctly, which leads to people becoming unwell. Sometimes this is as a result of an **infection** by harmful **micro-organisms**, which the body then has to get rid of.

Did you know?

Micro-organisms that cause disease are known as pathogens.

Figure 10.4 This person has a cold. It has been caused by a viral infection.

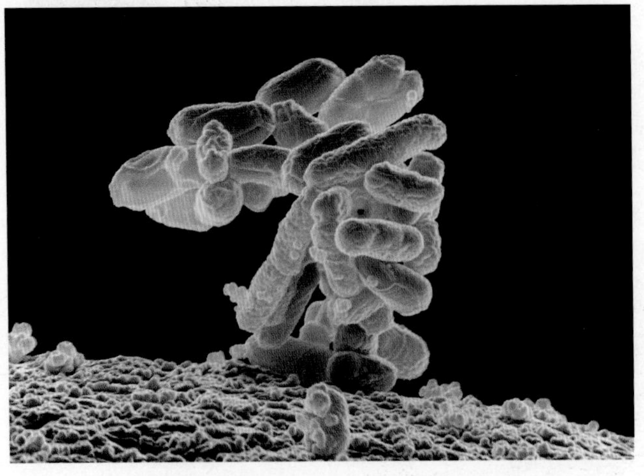

Figure 10.5 These **bacteria** are an example of micro-organisms. Sometimes micro-organisms can be useful to humans. Unfortunately, some are harmful and could potentially be life-threatening if they get inside the body.

Key words:

bacteria, complex, infection, micro-organism, organism

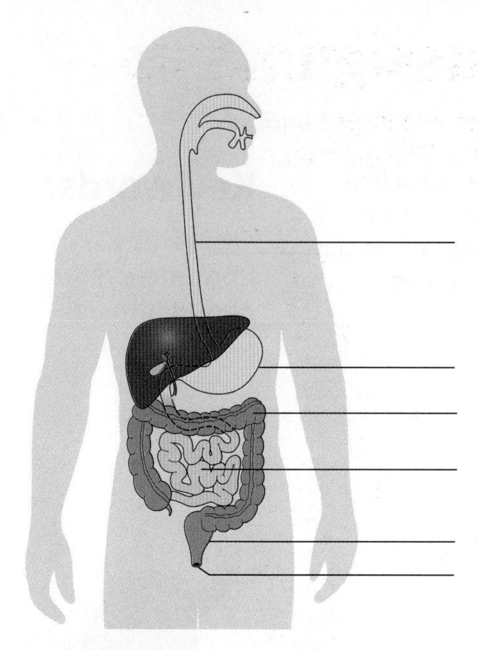

Figure 10.6 The main parts of the digestive system.

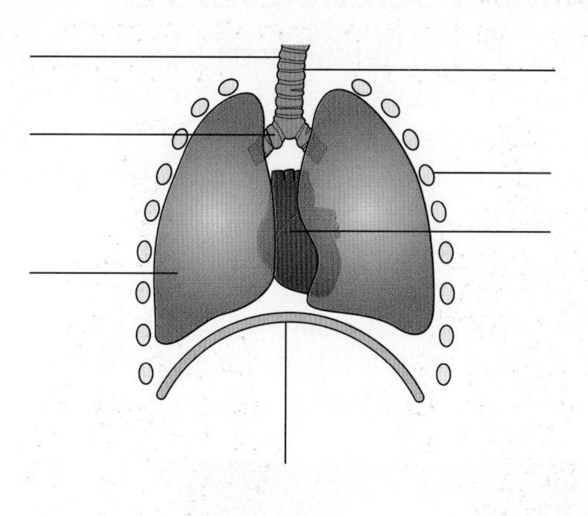

Figure 10.7 The main parts of the breathing system.

Staying healthy activities

1. **Copy the diagram and label the main parts of the digestive system shown in Figure 10.6 above.**
2. **Copy the diagram and label the main parts of the breathing system shown in Figure 10.7 above.**
3. **Find out what each of the main organs you have labelled do.**
4. **Make a list of ten different diseases. Find out which micro-organism each of these ten diseases is caused by (if any).**
5. **Not all micro-organisms are bad for you! Find out five ways that bacteria (or other micro-organisms) can be beneficial to humans.**

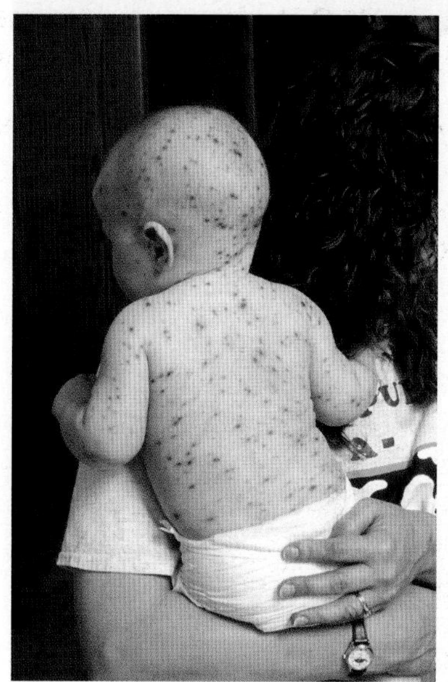

Figure 10.8 This baby has chicken pox, which is caused by the Varicella zoster virus.

Defence against attack

Essential science for P3

The body has to defend itself from attack every day. Micro-organisms like bacteria and viruses are all around us; if they successfully get around the body's defences, you may get ill. The body has an **immune system** that aims to get rid of any invading micro-organisms that do make it into the body.

Key words:

antibodies, immune system, phagocytosis

First lines of defence

The skin is a physical barrier that keeps micro-organisms out of the body. Where there is a break in the skin, the body has other methods to stop an invasion. Some of these defences are physical and some are chemical.

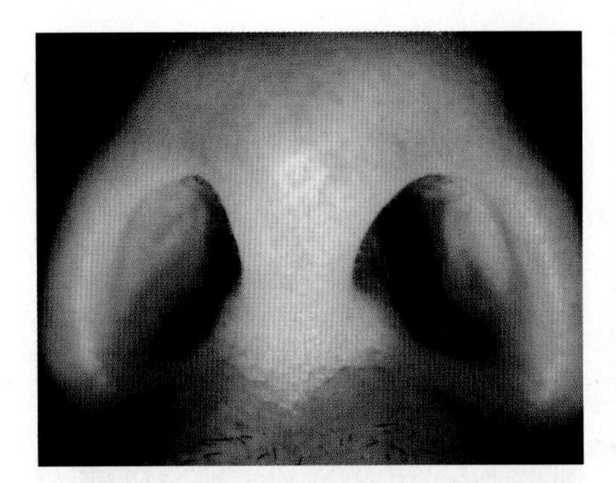

Figure 10.9 Nasal hairs trap dust and bacteria, preventing them from entering the lungs.

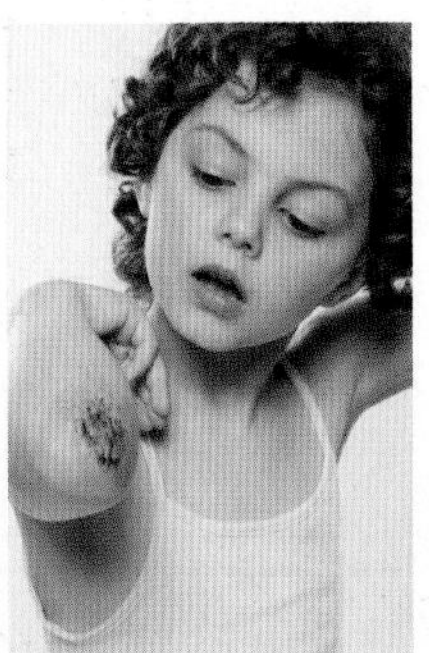

Figure 10.10 Blood platelets form a scab over a cut in the skin, preventing dirt and micro-organisms from entering the blood.

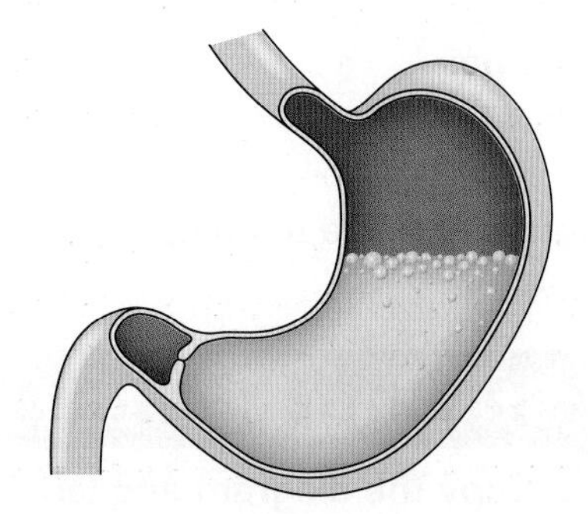

Figure 10.11 Stomach acid kills any bacteria that are swallowed with food or drink.

Next lines of defence

Sometimes micro-organisms do make it past the first lines of defence and cause an infection. White blood cells are then carried in the blood to the site of the infection. This often makes the area become inflamed (red and swollen), as there is more blood there than usual to bring lots of white blood cells to the area. Some white blood cells are able to 'eat' the invading micro-organisms! This process is known as **phagocytosis**. Other white blood cells produce chemicals called **antibodies** that attach to the micro-organisms and help to destroy them.

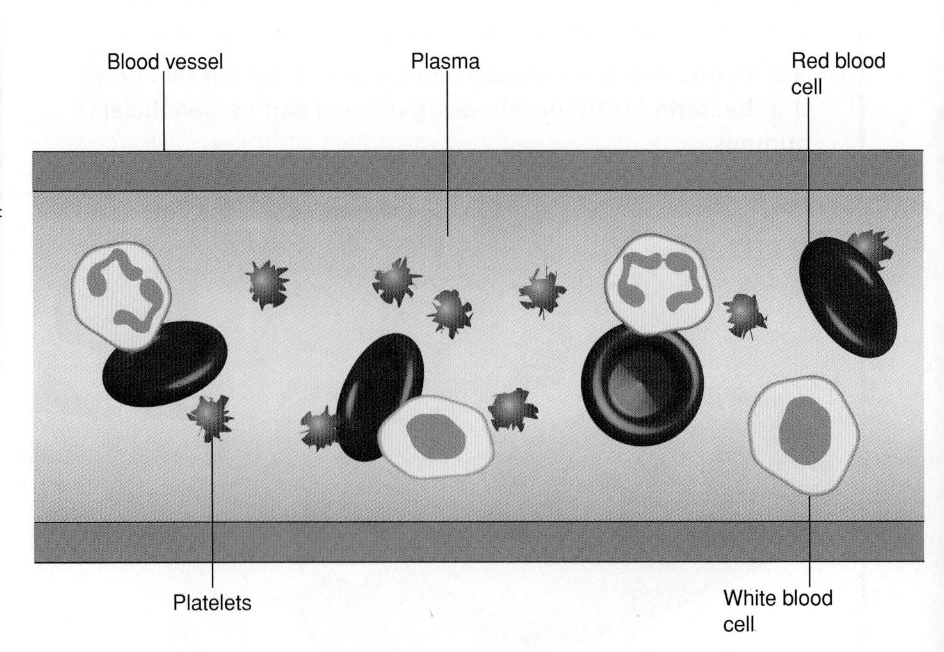

Figure 10.12 The blood contains white blood cells that help to protect the body from infection.

Task

You are going to create a presentation to give to teenage girls about how the body defends itself from attack by micro-organisms.

Your task:

Step 1: Research

Find out how the immune system defends the human body from attack by micro-organisms. You should find out about:

- **physical barriers**
- **chemical defences**
- **two things that white blood cells do.**

Find or create some diagrams and pictures to help you explain the points you make. Remember to reference the names of any websites that you collect useful diagrams or pictures from.

Step 2: Present

Use the information you have found and some of the useful diagrams or pictures to create a PowerPoint presentation that outlines how the immune system defends the human body.

Make the grade

P3 For P3 you must complete this task:

Outline how the immune system defends the human body.

Tip

A good presentation will have key points on it, rather than lots of writing.

Defending the body – how the immune system works

Essential science for M2

White blood cells

The white blood cell on the right is called a '**phagocyte**' (*phago* means eat, and *cyte* means cell). Phagocytes are able to 'eat' and destroy microscopic invaders. This is a '**non-specific**' response; the phagocyte cells will surround and destroy anything they find that is not meant to be inside the body.

The white blood cell below is called a **lymphocyte**. Lymphocytes are able to make and release chemicals called **antibodies**, which attach to bacteria and help make sure they are destroyed. This is a '**specific**' response; each antibody is made for a specific type of bacteria and will only attach to that one type.

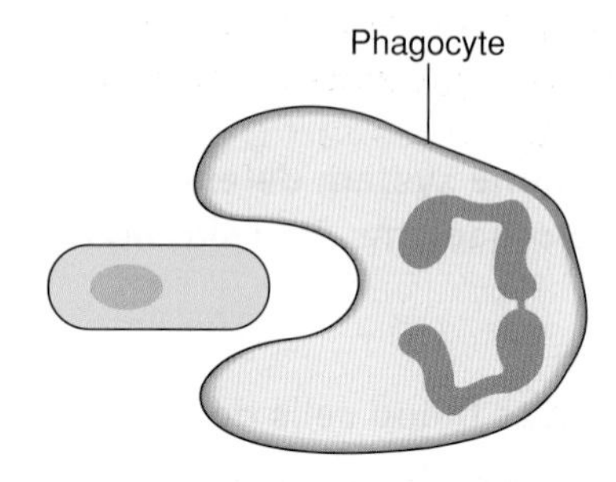

Figure 10.13 The white blood cell is 'eating' the bacterium. This is called phagocytosis.

Some micro-organisms release toxins (poisons) that make us ill. Lymphocytes can release specific **antitoxins** to protect the body from these poisons.

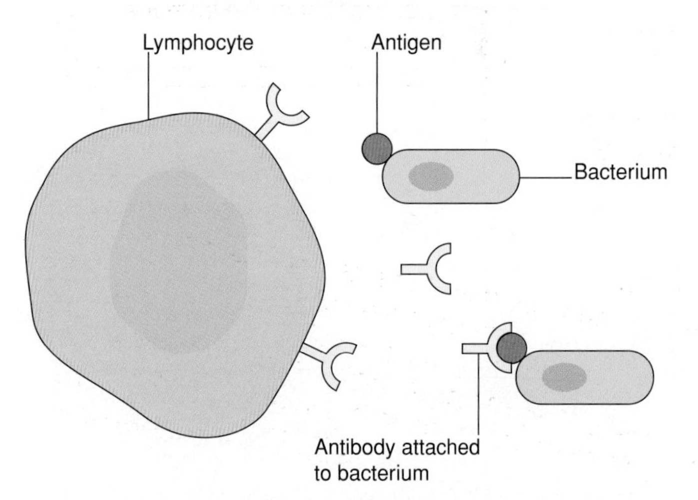

Figure 10.14 The lymphocyte makes an antibody that attaches to the **antigens** on the bacteria.

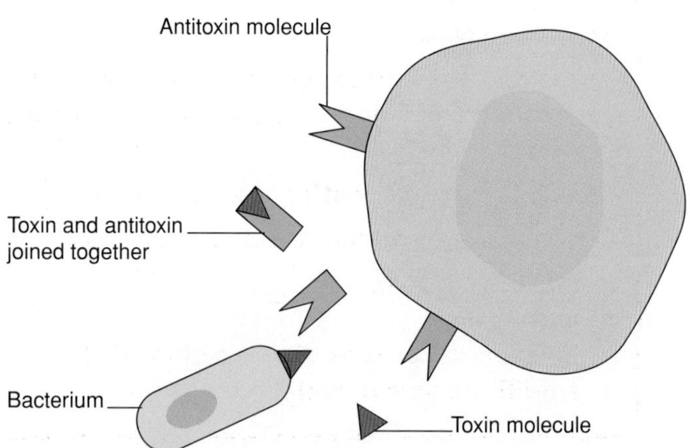

Figure 10.15 The lymphocyte makes an antitoxin that attaches to the toxin molecule and blocks its effect.

Did you know?

Antigens are molecules found on the surface of micro-organisms that the immune system can recognise and make antibodies against.

Key words:

antibody, antigen, antitoxin, lymphocyte, lysozyme, non-specific, phagocyte, specific

Protecting the eyes

The eye is a possible 'way in' for bacteria, so it has several features that help to protect it. Eyelashes are a physical barrier to help keep dirt out of the eye. Tears keep the eye moist and help to wash any dirt out of it. They also contain a chemical called **lysozyme** that kills any bacteria that do reach the eye.

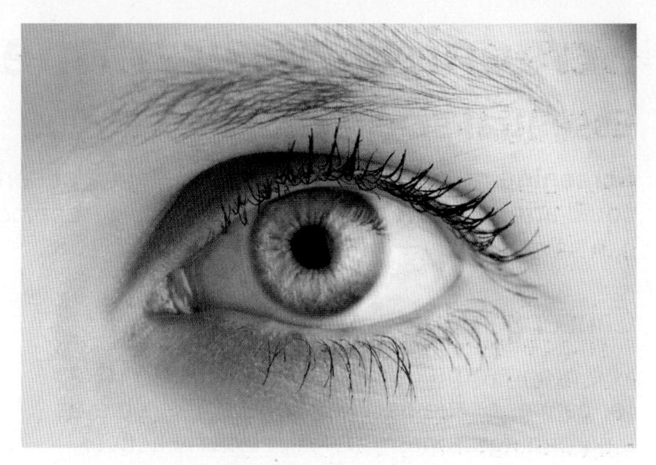

Figure 10.16 The eye has several adaptations that protect it from micro-organisms.

Task

You are going to develop some information sheets that you can give to teenage girls who have attended the presentation you made for P3. The information sheets must include a description of each component of the immune system.

Your task:

Step 1: Research

Return to the list of components of the immune system that you made for P3. Find out how each component works to protect the body.

Step 2: Create information sheets

Create four information sheets that cover:

- **physical barriers**
- **chemicals**
- **non-specific actions of white blood cells**
- **specific actions of white blood cells.**

Make the grade

M2 For M2, you must complete this task:

Describe the action of each component of the immune system.

Tip
Remember, these sheets are for teenage girls who may not be good at science. Make sure that you are using suitable language and describing clearly how each component of the immune system works.

Screening programmes

Essential science for P4

Screening programmes are used to detect abnormalities that could affect a person's health. Effective screening programmes can save lives.

Cancer screening

The programme of tests carried out to check for cancer of the **cervix** is a good example of a screening programme. All sexually active women should be regularly screened with a standard **smear test** every few years. Abnormal cells are identified by the smear test and can then be treated. This improves the chances of early diagnosis and saves lives.

Unfortunately, not all types of cancer can be detected using screening.

Antenatal screening

The health of an unborn baby can be checked using **antenatal** screening. Screening tests assess the risk of the baby being born with certain conditions, such as spina bifida or Down syndrome.

Newborn (postnatal) screening

A blood test is carried out on newborn babies to screen them for inherited disorders that may not be obvious to doctors. This is known as **postnatal** screening.

Vascular screening

Vascular screening is used to assess the risk of somebody having a stroke or heart attack. If it is detected early, vascular disease can be effectively treated.

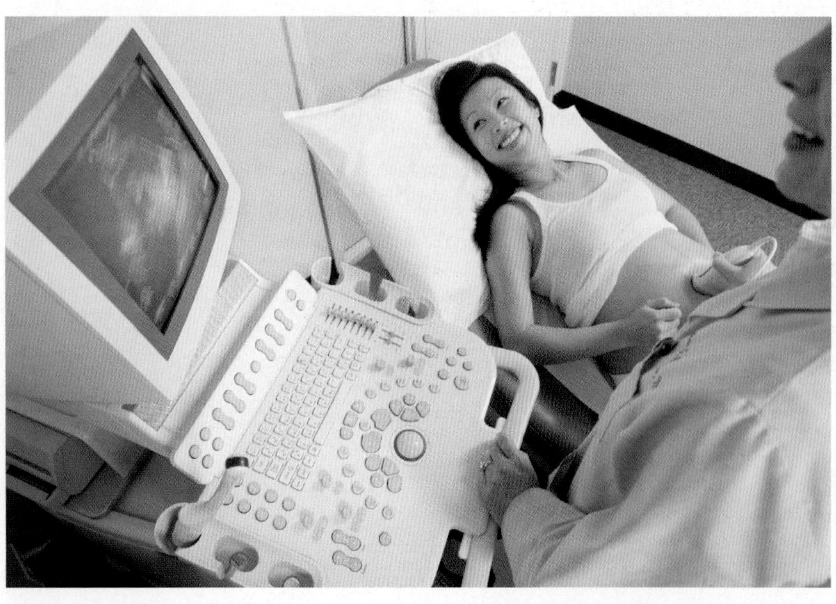

Figure 10.17 A screening test to check the health of an unborn baby.

Key words:

antenatal, cervix, postnatal, screening, smear test, vascular

Task

As part of your work for the Cancer Research charity, you are going to create a presentation to give to teenage girls about the importance of screening programmes. This will help to raise their awareness of cervical cancer and what can be done to prevent it.

Your task:

Step 1: Research

The main focus of your presentation is screening for cervical cancer, as you are trying to encourage the teenage girls to go for smear tests when they are older. In addition to this, you need to find out about one example of each of the other three types of screening (antenatal, newborn and vascular).

Find out:

- **what different screening programmes are available**
- **who they are available to**
- **what part they play in maintaining good health.**

Find or create some diagrams and pictures to help you explain the points you make in your presentation. Remember to reference the names of any websites that you collect useful diagrams or pictures from.

Step 2: Prepare

Now work on a series of powerful arguments about cervical cancer screening that will get through to teenage girls. Also find arguments about why people should bother with antenatal, newborn and vascular screening. You need ideas that will appeal to busy young people and ancourage them to take part in screening.

Step 3: Present

Use your information, including any useful diagrams or pictures, to create a PowerPoint presentation that outlines the importance of four health screening programmes.

Tip
Try to use a maximum of two PowerPoint slides for each of the four health screening programmes you cover.

Make the grade

P4 For P4, you must complete this task:

Identify the role of specific health screening programmes.

Vaccination

Essential science for M3

A **lymphocyte** is a white blood cell that can make and release **antibodies**. Different **pathogens** have specific antigens on their surfaces, which are like signatures. The body is able to make antibodies that can recognise specific antigens and stick to them. The first time that a pathogen enters the body, it can take some time for the body to make an antibody to attack it. However, once the body has met a pathogen and made an antibody once, it can quickly make the same antibody again. The person is now 'immune' to that pathogen.

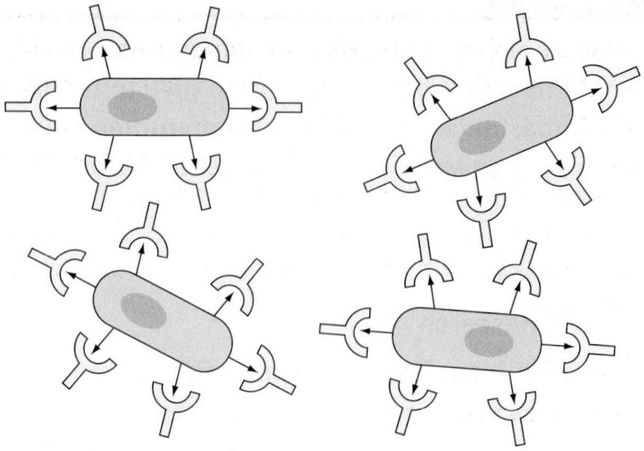

Figure 10.18 Antibodies binding to the antigens on the surface of a pathogen.

Vaccination

The diagram shows how a harmless version of a pathogen can be used to make the body immune to that pathogen in the future. This is known as a **vaccination**.

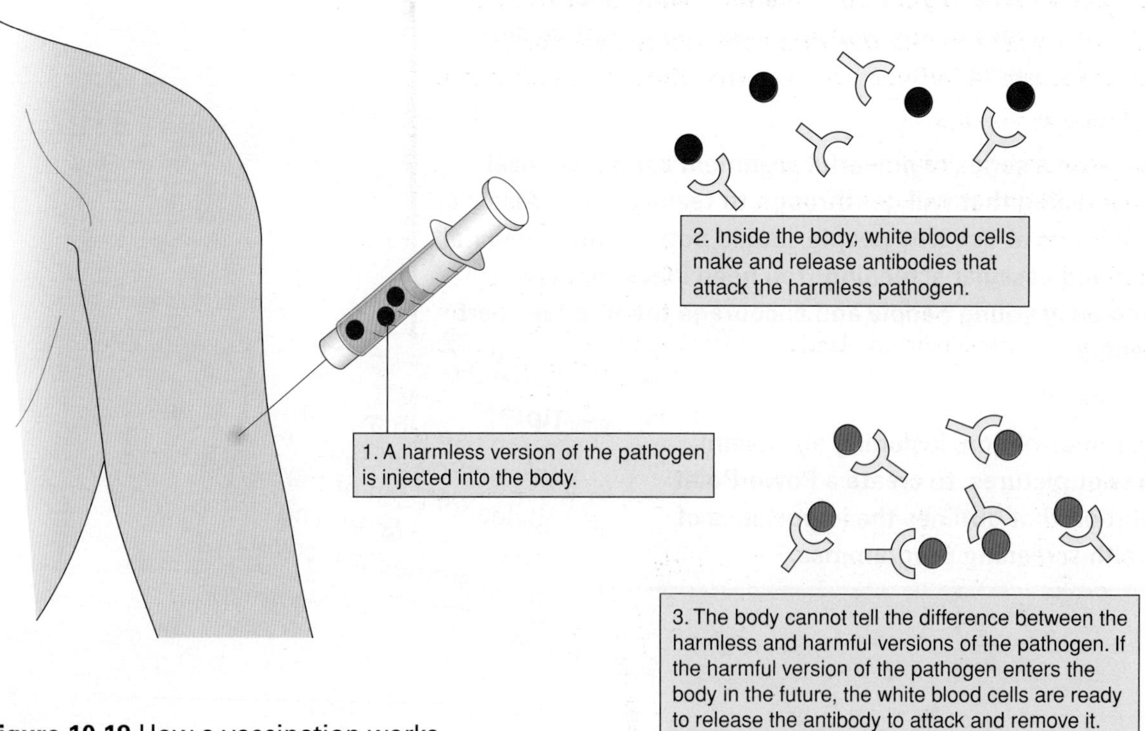

Figure 10.19 How a vaccination works.

Key words:

antibody, lymphocyte, pathogen, vaccination

Task

You are going to develop a leaflet about the HPV (human papilloma virus) vaccine, which is given to teenage girls. HPV is thought to cause cervical cancer. The leaflets will be given to teenage girls to raise awareness of why they should have the vaccination and how it affects the body. Your leaflet must include a detailed description of what happens after a vaccination is received, using clear diagrams to help you explain.

Your task:

Step 1: Research

Find out about the changes that occur in the body following the HPV vaccination. Your research should aim to answer the following questions:

- **What is HPV?**
- **What is a vaccine?**
- **What does the HPV vaccine contain?**
- **How does the vaccine work?**
- **What is meant by 'booster vaccinations'? Are they likely to be needed for the HPV vaccine?**
- **What side effects might there be after having the vaccine?**

Find some diagrams or pictures that help you to describe the changes that happen in the body following the HPV vaccination.

Step 2: Make the leaflet

Use your research to create your leaflet. It is a good idea to divide your leaflet into different sections, for example:

- **What is HPV?**
- **How does the vaccine work?**
- **What are the effects on the body?**

Tip
Make sure that you reference any websites you use.

Tip
Remember, the leaflet is for teenage girls who may not be good at science. Make sure that you are using suitable language and describing clearly how the immune system responds to a vaccination.

Make the grade

M3 For M3, you must complete this task:

Describe the changes in the human body following a vaccination.

How well do health programmes work?

Essential science for D2

Case study: Cervical cancer

Cervical cancer screening

- There are around 2800 new cases of cervical cancer diagnosed in the UK each year.
- The NHS Cervical Screening Programme was set up in 1988 by the Department of Health.
- Cervical screening can detect cell changes, but can't prevent infection in the first place.
- In the UK, cervical screening saves over 1000 lives every year and has prevented eight out of every ten cervical cancers from developing.
- Cervical screening in England is currently offered from the age of 25.
- Cervical screening can prevent around 75% of cancer cases in women who attend regularly.

HPV vaccination

- HPV vaccination in schools was introduced into the national immunisation programme in 2008, for girls aged 12–13.
- The HPV vaccine protects against the two strains of HPV (16 and 18) that cause cervical cancer in over 70% of cases.
- The vaccine is 99% effective in preventing the cervical abnormalities that can lead to cervical cancer caused by HPV types 16 and 18.
- Because the HPV vaccine does not protect against *all* cervical cancers, it is really important for all girls to have cervical screening, even if they have had the HPV vaccine.

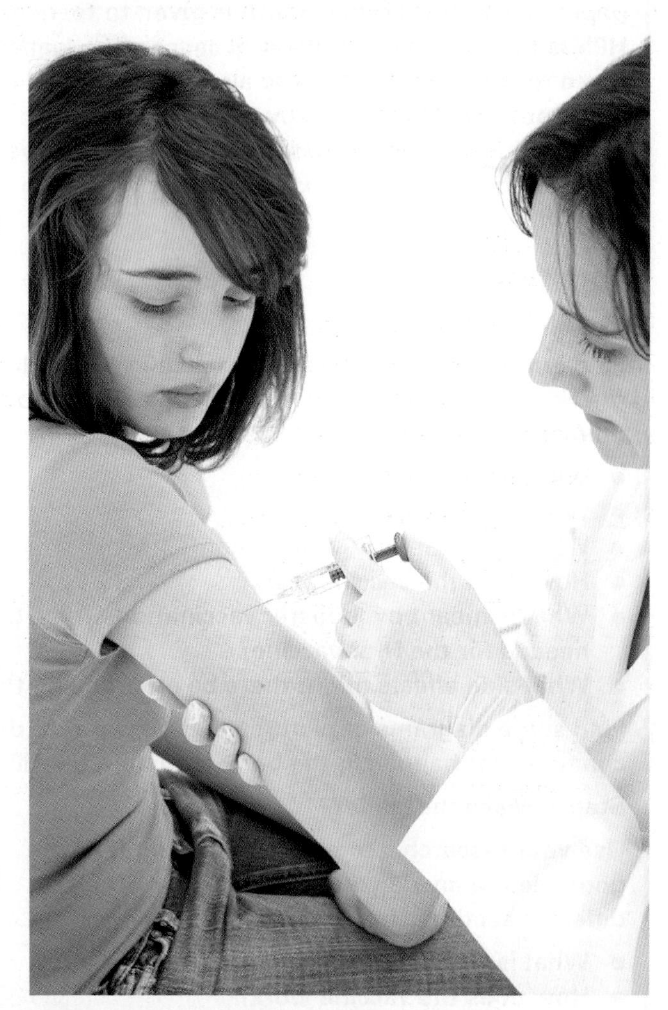

Figure 10.20 The HPV vaccine can help to protect against cervical cancer.

Task

There has been some resistance to the HPV vaccination, with some girls (or their parents) refusing the vaccination. In addition to this, some women do not attend cervical screening appointments regularly. Screening is currently not available automatically for women below the age of 25, and some people think that this is wrong and are campaigning for the age to be lowered.

Your task:

Step 1: Research

Use your findings from the earlier tasks and carry out further research to help you develop your evaluation. The following websites give lots of good information about cervical cancer, HPV, cervical screening and the HPV vaccine:

www.fightcervicalcancer.org.uk

www.immunisation.nhs.uk/Vaccines/HPV

Step 2: Develop an argument

Develop a balanced argument about the vaccination and screening programmes, which evaluates the effectiveness of both programmes. Your evidence must cover both advantages and disadvantages before an overall judgement on the effectiveness of the programmes is made.

Make the grade

D2 For D2, you must complete this task:

Evaluate the effectiveness of vaccination and screening programmes.

Unit 6

Chapter 11
Fighting disease

Hospital care

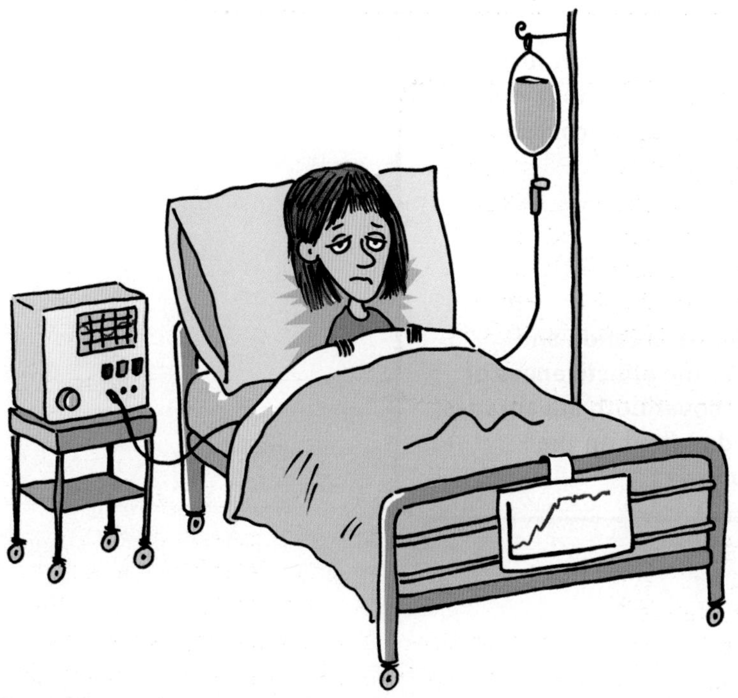

Figure 11.1

Scenario

People of all ages are treated in hospital for many different illnesses and conditions. There is a wide range of treatment options available for these different conditions. The diseases that people suffer from have many different causes. There are those you catch through micro-organisms entering your body. There are those you develop as you get older – like Crohn's disease, which usually begins to affect people when they are in their late teens. And there are diseases like cystic fibrosis that sufferers are born with.

You are a doctor in a hospital who has been asked to produce an illustrated report about the different ways that you treat illness and the procedures you use. You are required to investigate the effectiveness of different treatments, and to share your opinions on the use of different kinds of medical treatments.

Grading criteria for Fighting disease

To achieve a pass grade you need to:	To achieve a merit grade you also need to:	To achieve a distinction you also need to:
P5 carry out an investigation into the effects of antibiotics	**M4** using secondary data, carry out an investigation into the effectiveness of different kinds of medical treatment in the control of health	**D3** evaluate the use of different kinds of medical treatments, justifying your opinions
P6 describe what gene therapy is, giving examples of diseases and conditions associated with it		

Scenario

You work for a private healthcare company. You need to advertise the services that you can offer to your customers. The general public need to know what sort of healthcare they can expect if they subscribe to your health insurance scheme. You have been given the job of producing leaflets about how your particular sort of healthcare will work.

Keeping it local

- You could interview a local health worker about the variety of health conditions they deal with in their job and what they do on a day-to-day basis. You may be surprised at how varied their work can be.
- Use the *Yellow Pages* to find out about the range of medical services, doctors and medical jobs that there is in your area. The NHS is a huge employer.
- Carry out an interview with a family member, friend or classmate who has been in a local hospital for a long stay. Ask them to explain the range of health services and treatments they have seen in operation. Get them to comment on how well they were medically treated and if they were cured of their problem.

Did you know?

Did you know that in 1918 there was an outbreak of flu in Spain? Within months it had spread all over the world and killed more than 21 million people. When a disease spreads across many countries it is called a pandemic. When it affects a large number of people it is called an epidemic.

Careers

- Microbiologist
- Pathologist
- Pharmaceutical technician
- Clinical trial administrator
- Nurse
- Social worker
- Medical laboratory technician
- Medical research worker

Background science for the assignments

Infectious disease

The body can be infected by **pathogenic** (disease-causing) micro-organisms including bacteria, viruses and fungi.

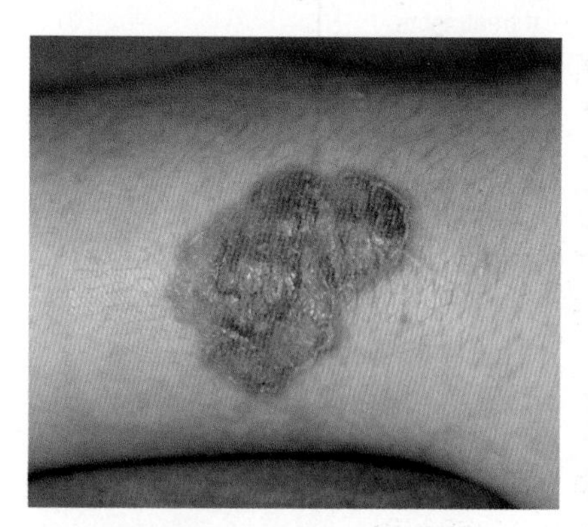

Figure 11.2 Impetigo is caused by a bacterial **infection**.

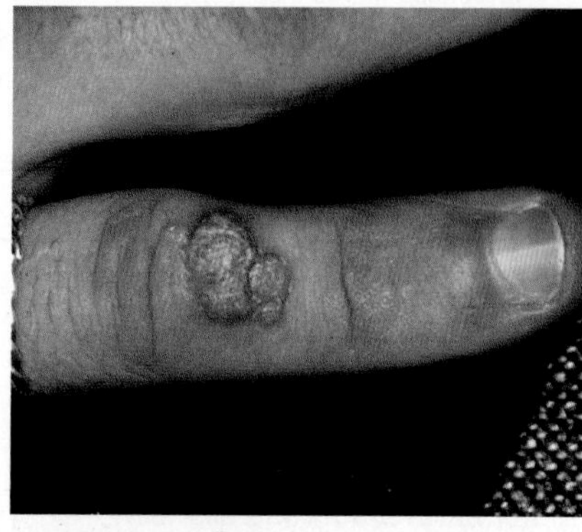

Figure 11.3 Warts are caused by a viral infection.

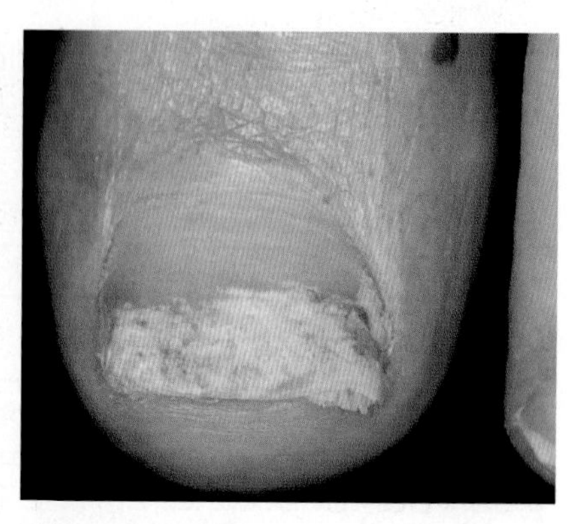

Figure 11.4 This toenail has a fungal infection.

Inherited disease

Babies can be born suffering from genetic disorders or diseases that they have **inherited** from their parents.

Chromosomes contain **genes**. All genes have **alleles**; different versions of the gene. Inherited disorders are caused by faulty alleles.

Cystic fibrosis is caused by a faulty allele that is **recessive**. This means you have to inherit a faulty allele from both your mother and your father in order to have symptoms of the disease.

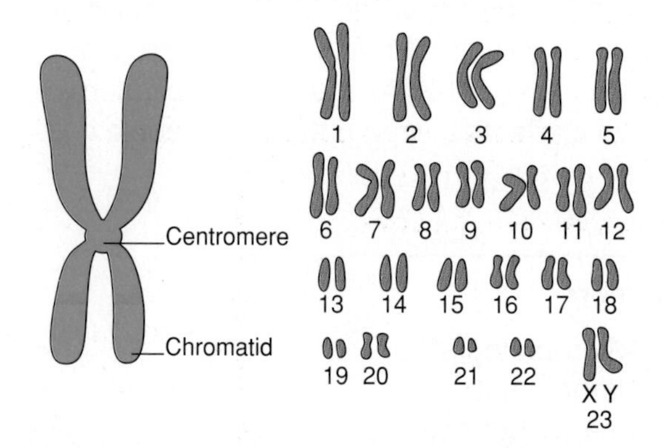

Figure 11.5 Chromosomes are long pieces of **DNA** found in the nucleus of a cell. There are 23 pairs of chromosomes in every human cell. DNA is considered to be the blueprint for the human body; it contains all our genes.

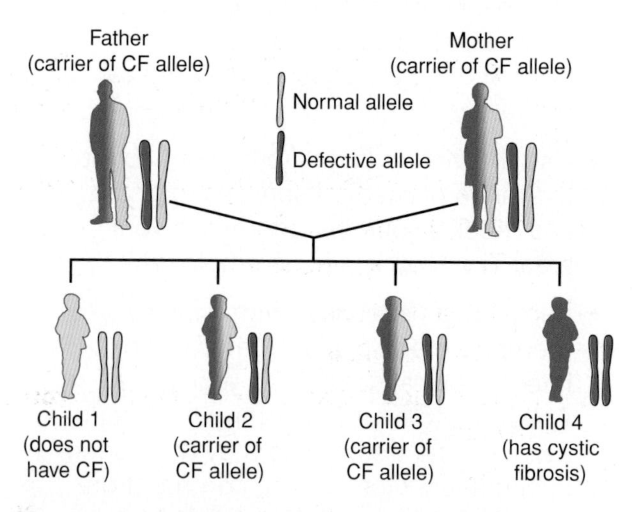

Figure 11.6 Inheritance of cystic fibrosis.

Some inherited diseases are caused by a faulty allele that is **dominant**. This means that you only need to inherit the faulty allele from one of your parents in order to have symptoms of the disease.

Key words:

allele, bacteria, chromosome, DNA, dominant, fungi, gene, infection, inherit, pathogenic, recessive, virus

Medical treatments

Medical treatments are used to treat all kinds of diseases. The type of treatment used depends upon the illness.

Antibiotics

The first antibiotic, penicillin, was discovered by Alexander Fleming in 1928 but was not used on patients until 1940. Antibiotics can be obtained from moulds and microbes or can be made using chemical reactions in laboratories. They can be used to treat many different diseases, but overusing antibiotics has led to some bacteria becoming resistant to them.

Gene therapy

This is a new type of medical treatment, which can provide relief from the symptoms caused by some genetic diseases. It is most commonly used to treat cystic fibrosis and haemophilia.

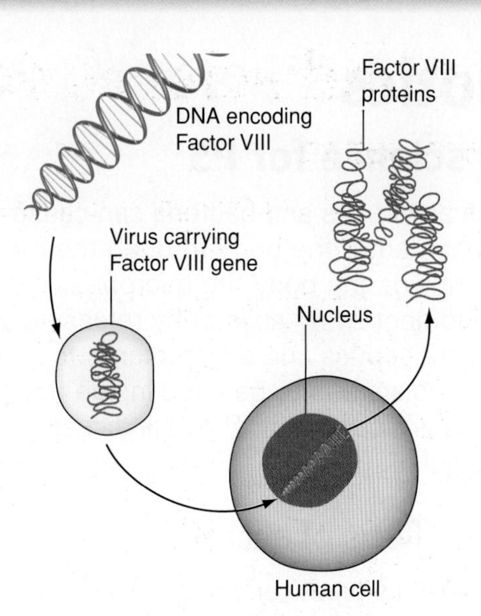

Figure 11.7 Gene therapy can be used to treat genetic diseases.

Blood grouping and transfusion

It is important to know your blood group if you need to have a blood transfusion or if you are pregnant. It can be determined from a simple blood test.

Stem cells

Stem cells can now be grown and transformed into many types of specialised cells and tissue, such as muscle and nerve cells, using a technique called tissue culture.

Fighting disease activities

Infectious disease

1 **For each of these diseases: athlete's foot, common cold, cholera; find out:**
- **a)** **what causes it**
- **b)** **what the symptoms are**
- **c)** **how it is treated**
- **d)** **how common it is**
- **e)** **if you have the disease once, whether you become immune to future infections.**

Inherited disease

2 **Carry out some research into cystic fibrosis. What does it mean to be a 'carrier'? How likely is it that a child will be born with cystic fibrosis if one of its parents is a carrier?**

3 **Find the name of an inherited disorder that is caused by a dominant faulty allele. Do some research about it. Can you be a carrier? How likely is it for a child to be born with the disorder if one parent has the faulty gene?**

Antibiotics

Essential science for P5

Key words:

agar plate, antibiotic, inhibit, MRSA, zone of inhibition

Microbes such as viruses and bacteria can cause many different types of disease. They can enter the body through the mouth or nose or through cuts and wounds. Once inside the body the microbes can start to multiply and cause harm by attacking and destroying cells or by releasing poisons into the bloodstream.

We can use antiseptics and disinfectants to kill bacteria outside the body, but they are far too poisonous to be used inside the body – they would kill us! This is why **antibiotics** are so useful. They kill disease-causing bacteria inside the body without killing us!

Practical: The effects of antibiotics

A teacher carried this practical out as a demonstration for her pupils, as working with bacteria can be hazardous.

- An **agar plate** was prepared with bacteria.
- Antibiotic-impregnated discs were placed on the agar.
- The agar plate was sealed and incubated for 24 hours.
- The antibiotics diffused from the discs into the agar and **inhibited** the growth of bacteria around them.
- Clear areas (**zones of inhibition**) could be seen where the bacteria had been killed.
- The diameters of the clear zones were measured.
- This allowed the effectiveness of the different antibiotics to be compared.

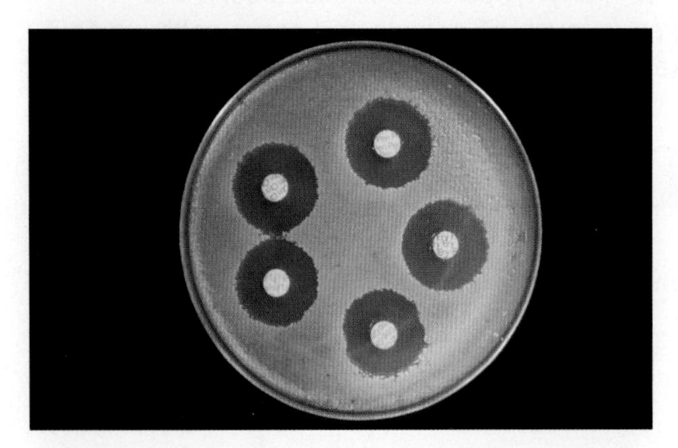

Figure 11.8 An agar plate with bacteria growing on it and antibiotic discs placed on the agar.

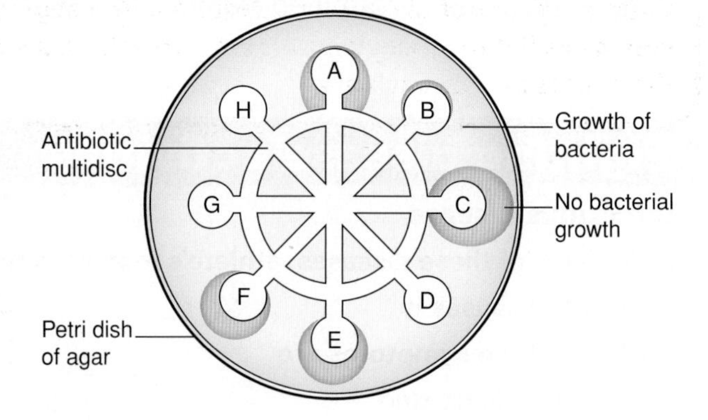

Figure 11.9 The effect of antibiotics A–H on the bacteria.

Figure 11.9 shows the results of the investigation.

Antibiotic resistance

Overuse of antibiotics has led to problems; bacteria have evolved through natural selection to be resistant to the antibiotics. **MRSA** (methicillin-resistant *Staphylococcus aureus*) is a well-known example of a bacterium that causes dangerous infections in many hospitals.

Questions

1. Which antibiotics were able to inhibit the growth of the bacteria?
2. Which antibiotic was the most effective?

Task

As part of your research for the hospital you need to test the effects of some antibiotics on bacteria.

- **You will be provided with agar plates containing bacteria.**
- **You will be provided with antibiotic discs or antibiotic multi-discs.**
- **You must follow the safety precautions!**

Your task:

Step 1: Predict

Make a prediction. Which antibiotic do you think will be the best at inhibiting the growth of the bacteria?

Step 2: Investigate

Carry out the investigation (or watch the demonstration) to see which antibiotics work well to reduce the growth of the bacteria.

Step 3: Results

Answer the following questions:

- **What did you find out?**
- **Which antibiotic was most effective?**
- **Was your prediction correct?**

Draw a diagram to represent the results.

Step 4: Present

Write a full report of your investigation. You should include an aim, a prediction, a description of the method, a summary of the results and a conclusion.

Make the grade

P5 For P5, you must complete this task:

Carry out an investigation into the effects of antibiotics.

Gene therapy

Essential science for P6

Gene therapy is an experimental technique that uses healthy versions of genes to treat or prevent a disease.

Cystic fibrosis (CF) is a genetic disease that occurs when a faulty gene is inherited from both parents. CF causes the body to produce thick, sticky mucus that clogs the lungs and leads to infection. It also blocks the pancreas, which stops digestive enzymes from reaching the intestine where they are needed in order to digest food. Symptoms include poor growth and damage to the lungs, pancreas and liver. People with cystic fibrosis have a life expectancy of about 31 years. However, with more advances in the treatment of this disease, babies born with CF these days are expected to live longer. More than 8000 people in Britain suffer from cystic fibrosis.

Have a look at the Cystic Fibrosis Trust website for further information about this life-threatening disease.

Gene therapy has been developed to help treat cystic fibrosis. A normal copy of the healthy gene is inserted into the lung cells that are producing the thick mucus so that they start to work correctly. This type of genetics has only been developed in recent years. This makes gene therapy a relatively new treatment. About ten years of detailed scientific research and development is needed before the actual clinical trials can be started to investigate gene therapy for different diseases such as sickle cell anaemia, inherited blindness and blood disorders. There is a lot of research going on in these areas.

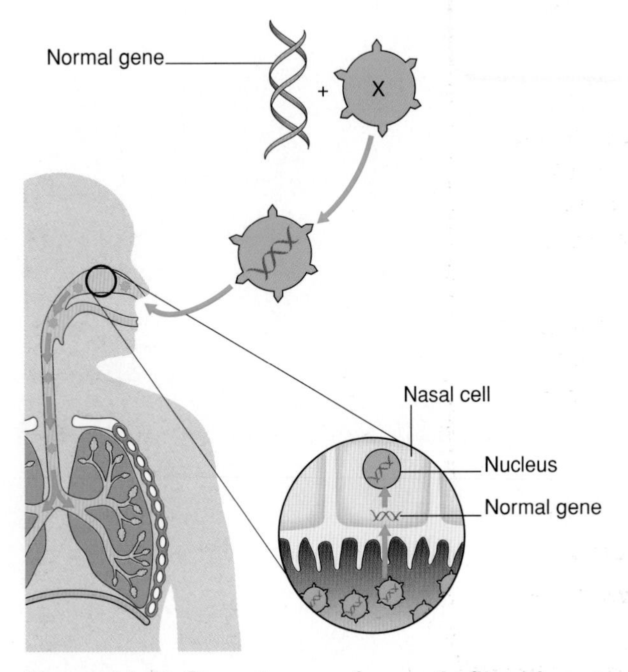

Figure 11.10 Gene therapy for cystic fibrosis involves inserting normal, healthy genes into the sufferer's lungs.

Questions

1 Research gene therapy.
   a) Which diseases can gene therapy be used to treat?
   b) What are some of the problems surrounding gene therapy treatment?

Key words:

gene therapy

More about gene therapy

Medics expect gene therapy will be used to treat a wide range of diseases, not just cystic fibrosis. Examples are:

- FH – a disease that clogs arteries with fat causing heart attacks. Part of the patient's liver would be removed, the genetic material inserted, then the liver put back.
- Aids – Acquired Immunodeficiency Syndrome. Genes could be inserted in the blood that stop the Human Immunodeficiency Virus from replicating.
- Cancers – Doctors could carry out gene therapy on some of the patient's own cancer cells and put them back in, making all the cancer cells die or become susceptible to common drugs.
- Gaucher's disease – makes your bones fall apart. Gene therapy could make cells produce the missing enzyme to prevent this.

Task

You are a genetic counsellor who has to talk to families about gene therapy and how it can be used to treat certain genetic disorders.

A couple have a son with a genetic disease and have come to you for some information. You need to tell them about the symptoms associated with the disease. You will also need to describe what gene therapy is using pictures and diagrams and explain how it can be used to treat the disease.

Your task:

Step 1: Describe the disease

Prepare some information explaining what the disease being treated is, how it is inherited and what the symptoms of the disease are. You should also give some examples of other genetic diseases that can be treated using gene therapy.

Step 2: Gene therapy

Describe what gene therapy is and how it can be used to treat the symptoms of the disease you have chosen. You should include pictures and diagrams to help you describe the process clearly.

Step 3: Present

Prepare a PowerPoint presentation to show to the family who want to know more about gene therapy and the disease.

Alternatively, you could prepare a series of posters with the information on them.

Make the grade

P6 For P6, you must complete this task:

Describe what gene therapy is, giving examples of diseases and conditions associated with it.

Medical treatments

Essential science for M4

Antibiotics

Antibiotics are very effective in the treatment of bacterial infections and have saved millions of lives since their introduction in 1940. They can be used to treat bacterial infections but not viral infections, because viruses reproduce inside the cells of the body.

Hospitals treat many infections and use huge amounts of antibiotics, and as a result of this some bacteria have become resistant to antibiotics. This means that they are much harder to kill. For example, you may have heard of MRSA (methicillin-resistant *Staphylococcus aureus*).

Antibiotics are relatively cheap to produce and can be made on a huge industrial scale.

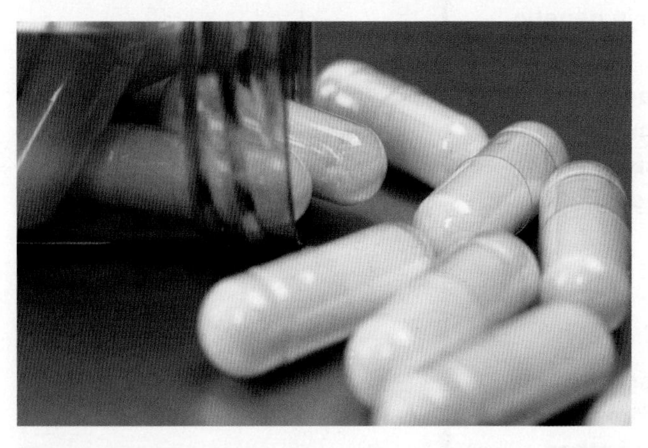

Figure 11.11 Antibiotics have saved millions of lives.

Gene therapy

Gene therapy is a relatively new treatment used to treat people with certain genetic diseases. You will have learnt about the genetic disease cystic fibrosis already when you completed P6 (see page 114).

Question

Stem cells can be:
totipotent, pluripotent, multipotent, oligopotent or unipotent.
Find out the meanings of these words.

Blood transfusions

If a patient loses a lot of blood, for example after an accident, during surgery or during childbirth, they might need a **blood transfusion**. This is where the blood they have lost is replaced with blood from a donor. Blood banks provide a transfusion service for blood and other blood products such as plasma. All blood products must be prescribed by a medical doctor, as it is vitally important that the correct blood for your blood group is used. If you received the wrong blood type, you could suffer an acute haemolytic reaction. This could lead to renal failure and even death.

Did you know?

You can check the national current blood stocks on the **National Blood Service** website. Which blood groups have the most stock and which have the least?

Stem cells

Stem cells can be found in all multicellular organisms. There are two types of stem cells: embryonic and adult. Both types have the ability to renew themselves and change into different kinds of specialised cells through cell division (mitosis), although embryonic stem cells are more versatile. Stem cell research began in the 1960s. Stem cells can now be grown and transformed into many types of specialised cells and tissues, such as muscles and nerve cells, using a technique called cell culture.

Figure 11.12 Stem cells with a fluorescent green marker.

Key words:

antibiotics, blood transfusion, gene therapy, National Blood Service, stem cells

Task

As part of your ongoing research for the hospital you work for, you need to find out more information about the different medical treatments you have studied. You will need to collect data and evidence to back up your research. When you have collected the information you will be able to say how effective each type of treatment is in maintaining good health.

Your task:

Step 1: Choose

Choose two types of medical treatment that you have been learning about. You can pick two from this list:

- **the use of antibiotics**
- **gene therapy**
- **blood transfusions**
- **the use of stem cells.**

Step 2: Research

For both of your treatments, collect information about them to show how effective each is. As part of your investigation, try to answer some of the following questions:

- **When was the treatment developed?**
- **How often is the treatment used?**
- **How many people are treated in the UK each year using this treatment?**
- **Are there any problems with the treatment?**
- **How effective is the treatment?**

Step 3: Present

Present your investigation in the form of two reports, which should be titled:

'The effectiveness of (named medical treatment) in the control of health'

Make the grade

M4 For M4, you must complete this task:

Using secondary data, carry out an investigation into the effectiveness of different kinds of medical treatment in the control of health.

Evaluate the medical treatments

Essential science for D3

In this section you will need to comment on the advantages and disadvantages of the different kinds of medical treatments and then evaluate their use, justifying your opinions. You could use a table like the one below to help you get started.

Medical treatments	Advantages	Disadvantages
use of antibiotics		
gene therapy		
blood transfusions		
use of stem cells		

When looking at advantages and disadvantages, think about these factors:

Cost:

- Is there a huge cost to the NHS of using the treatment?
- What was the original cost of developing the treatment?

Time:

- How long did the research and development take?
- How long did the clinical trials take?

Availability:

- How easily is the medical treatment available?
- Is there a waiting list for the treatment?
- What if someone has a very rare blood group, for example?

Ethical issues:

- Are there any ethical issues surrounding the treatment or the research used to develop it?
- For example, is it ethical to use embryos for research?

Figure 11.13 Many people object to the use of embryos for research.

Religion:

- Will any religions object to the treatment?
- For example, some religions do not agree with blood transfusions or the use of stem cells from embryos.

Success:

- How successful is the treatment?
- How has it changed the lives of the people treated?

Consequences of misuse:

- Are there any problems associated with the misuse of the treatment?
- For example, some bacteria have become resistant to antibiotics, leading to hospital-acquired infections (MRSA).

Task

You are continuing with your research for the hospital. You need to evaluate the uses of the four different kinds of medical treatment you have been studying. You can add this research to the report that you have already completed for M4.

Your task:

Step 1: Advantages and disadvantages

For each of the four medical treatments, outline the advantages and disadvantages of each one. You may want to do this in a table format. You need to consider factors such as cost, time, availability, ethics, religion, success, and consequences of misuse.

Step 2: Evaluate

Now that you have looked at the advantages and disadvantages of all four medical treatments, you need to evaluate each one. Justify your judgements using evidence that you have found in your research.

Step 3: Present

Present your evaluation as a report. You can do this as a separate report or add it to the report that you completed for M4.

Make the grade

D3 For D3, you must complete this task:

Evaluate the use of different kinds of medical treatments, justifying your opinions.

Index

Acknowledgements

The Publishers would like to thank the following for permission to reproduce copyright material:

Photo credits
p.8 © demarco – Fotolia.com; **p.12** *l* Rex Features, *r* Geoff Moore/Rex Features; **p.13** *t* NASA, *m* NASA/Kim Shiflett, *b* Comstock/photolibrary.com; **p.14** *l* Charles D. Winters/Science Photo Library, *r* © Siede Preis/Photodisc/Getty Images; **p.15** *l* Charles D. Winters/Science Photo Library, *r* © sciencephotos/Alamy; **p.16** © Martin Shields/Alamy; **p.22** *tl* Brooke Fasani/Photonica/Getty Images, *tm* © Diana Mastepanova – Fotolia.com, *tr* Nigel Monckton – Fotolia, *br* Leonard Lessin/Science Photo Library, *bmr* © Liv Friis-larsen – Fotolia, *bml* Brooke Fuller – Fotolia, *bl* © Iain Masterton/Alamy; **p.28** © 1997 John A. Rizzo/PhotoDisc/Getty Images; **p.29** *l* adisa – Fotolia, *r* ZTS – Fotolia; **p.30** *t* © F. Bettex – Mysterra.org/ Alamy, *r* © Biosphoto/Pierre Huguet/Still Pictures; **p.32** *l* Rebecca Naden/PA Images, *r* LIU JIN/AFP/Getty Images; **p.33** Caroline von Tuempling/ Iconica/Getty Images; **p.34** *tl* Dmitry Nikolaev – Fotolia, *ml* © Robert Destefano/Alamy, *bl* © Stockbyte/Getty Images, *bm* SSPL/Science Museum/Getty Images, *br* Martyn F. Chillmaid/Science Photo Library; **p.36** *l* © Purestock/Ingram Publishing/Getty Images, *m* David McCarthy/ Science Photo, *r* diter – Fotolia; **p.37** © Stockbyte/Getty Images; **p.38** USMC/Science Photo Library; **p.40** Patrick Dumas/Look At Sciences/ Science Photo Library; **p.42** © Alchemy/Alamy; **p.43** © Corbis. All Rights Reserved.; **p.44** © Imagestate Media; **p.45** Alekss – Fotolia; **p.46** *t* © Alchemy/Alamy, *b* Adam Hart-Davis/Science Photo Library; **p.50** © Blade_kostas/istockphoto.com; **p.52** Konovalikov Andrey/ istockphoto.com; **p.53** Peter Dazeley/The Image Bank/Getty Images; **p.54** *tl* M. Meliksetyan – Fotolia, *tr* Vladimir Shevelev – Fotolia, *b* Roman Milert – Fotolia; **p.55** Nigel Monckton – Fotolia; **p.57** © Cindy Haggerty – Fotolia.com; **p.58** © JCPJR/istockphoto.com; **p.60** © S. Meltzer/ PhotoLink/Photodisc/Getty Images; **p.61** © Greg Epperson/fotolia.com; **p.62** © Images of Birmingham Premium/Alamy; **p.64** soupstock – Fotolia; **p.70** © Compix/Alamy; **p.72** *l* Chedges – Fotolia, *r* M&M – Fotolia; **p.73** Alexander Raths – Fotolia; **p.82** *l* © thierry burot – Fotolia, *r* © Robert Stainforth/Alamy; **p.83** © Imagestate Media; **p.86** © Martin Shields/Alamy; **p.94** *t* Crown Copyright/Courtesy Department of Health, *l* Chris Radburn/PA Wire/Press Association Images, *r* Sam Ogden/Science Photo Library; **p.95** © Image Source/Alamy; **p.96** *l* © Imagestate Media, *r* Photo by Eric Erbe; digital colorization by Christopher Pooley/Material produced by ARS is in the public domain; **p.97** *l* © Ingram Publishing Limited, *m* © Ingram Publishing Limited, *r* Lowell Georgia/Science Photo Library; **p.98** *l* © Maximilian Weinzierl/Alamy, *r* © altrendo images/Stockbyte/Getty Images; **p.101** © m.arc – Fotolia; **p.102** Katrina Wittkamp/The Image Bank/Getty Images; **p.106** May/Science Photo Library; **p.110** *l* Dr P. Marazzi/Science Photo Library, *m* CNRI/Science Photo Library, *r* Dr P. Marazzi/Science Photo Library; **p.112** CNRI/Science Photo Library; **p.116** *l* © Imagestate Media, *r* Science Photo Library; **p.118** © Creatas/Comstock/Photolibrary Group Ltd.

Every effort has been made to trace all copyright holders but if any have been inadvertantly overlooked the Publishers will be pleased to make the necessary arrangements at the first opportunity.